enjoy this

Inspiring memoir

[illegible], MD

No Dream Beyond My Reach

One woman's remarkable journey from Cambodian refugee to American MD

Sopheap Ly, MD

AuthorHouse™
1663 Liberty Drive
Bloomington, IN 47403
www.authorhouse.com
Phone: 1-800-839-8640

First published by AuthorHouse 6/10/2009

ISBN: 978-1-4389-8457-5 (e)
ISBN: 978-1-4389-8456-8 (sc)

Library of Congress Control Number: 2009904670

Printed in the United States of America
Bloomington, Indiana

This book is printed on acid-free paper.

Dedication

To my beloved and extraordinary dad for planting my early dream to succeed in the medical profession, for giving me excellent genes with strong mental and physical capacities, for instilling in me a can-do attitude, and for fueling me with countless motivations. I wish you were alive to see me excel.

To my husband, Kaustubh, for being such a loving and understanding man who has so much confidence in me as a loving, busy, and ambitious wife.

And to our precious twin girls (twice the love), Sonali and Manali, for being happy and healthy and for giving us so many reasons to live well and have fun. Mommy and Daddy love you very much and hope to give you every opportunity to shape your own dreams. We hope that we are good role models for you.

Table of Contents

Acknowledgments

Be careful of the company you keep. Good company will build your character; the other kind won't. I especially want to thank the following individuals who have certainly been good company!

I would like to thank all the medical students at Howard University's College of Medicine for their many hours of support emotionally and spiritually—especially the sophomores who frequently came to sing songs and pray for us to pass our exams. There was one sophomore medical student who had a voice like an angel, and she blessed us with her sweet songs.

Thank you to Kevin and Hans. You were strong voices of encouragement during our late nights of study. Your positive example of staying up late and going the extra mile was a real inspiration for me to stay the course. I would always make a u-turn after seeing you still studying in the early hours of the morning. I would think: *If Kevin and*

Han can do it, I can do it too! Congratulations, Hans, for getting accepted for residency at the Mayo Clinic.

I want to thank my professors, teachers, mentors, and friends in medicine, especially Gary Lask, MD, at UCLA, who taught me the value of never giving up and to always work hard to pursue my dreams.

I would like to thank all of those who inspired and cheered me on when I was discouraged and didn't know if I could reach my dream. You gave me light when I was blind. You gave me wisdom and knowledge when I was foolish and ignorant. You provided resources when I had none. Special thanks to Dr. Mijeung Gwak for giving me food when I was hungry and helping me out when I was broke.

I want to thank three very special Cambodian doctors: Morakod Lim, MD, Sarady Tan, MD, and Linath Lim, MD. You gave countless hours encouraging me to stay focused and not give up. You were an inspiration.

Delight Satter, MPH, and Luis Pena, PhD, you have been true friends, believing in me during tough times. Special thanks to you, Delight, for visiting me while I was in medical school and for offering me a great place to stay temporarily during my rotation at UCLA Medical Center.

Thanks to Ms. Silvina Camacho Pichardo for all the wise advice you have given me throughout my life. You are an awesome role model. Thanks for giving me the big idea to translate this book into Spanish and finding me a special Spanish translator. Since I can read Spanish moderately well, I can't wait to read my own book in Spanish!

Thanks to Anna Marie Rodriguez, MD, and Rodanthi C. Kitridou, MD, FACP, MACR, Professor Emerita of Medicine (Rheumatology), USC Keck School of Medicine, for your unselfish love and caring toward my family. I am so fortunate to have special friends like you from LAC-USC Medical Center. I know you love to travel to Cambodia, and I hope my book will enhance your understanding of Cambodia even more.

I would like to thank my friends Grace Hee Kang Lee, DDS, and her husband Sam Lee, DDS, for the kindest hearts. You have been there for me through many of my hardships, from studying for the medical board exams to the frequent visits while I was pregnant with my twins. Grace, I will never forget you for the raw and pure love you have toward my family. And I will make sure my twins remember that you are the one who brought me all the special foods day after day, so that I could produce milk for them. I want to thank your little daughter Rebecca and little son Thomas, for also giving their love to my twin girls.

Thank you, Dr. Katherine Tran and your husband, Phillip Richardson, MD, for caring enough to help me understand more about life and relationships. Thank you Aretha Makia, MD, and Asek Nelson Makia, MD, for giving me so much love and understanding. Thank you for your kindness. I'll never forget you, and I'll never forget the times we shared laughing and crying as we struggled to get through medical school.

I would like to thank my mother-in-law, Surekha Reva Vinze Marathe, who was my wonderful friend for more than ten years before I married her son. You have since passed away, but I will always remember your generosity. You advised me to donate money to charity when I finished medical school. It is my honor to be able to fulfill your wishes.

To cousin Manasee Marathe, MBA, thanks so much for all your support and love always. You, your husband, Ameet, and your brother, Omkar Marathe, MD, were the first ones to arrive at the birth of our twins after camping out almost all night at the hospital to welcome them. This is what Aunty Manasee wrote on behalf of the twins:

> *To Our Beautiful Mommy. You overcame every obstacle to make sure we came into this world healthy and happy. We will forever be grateful to our Mommy for all she did and all she does to make our lives wonderful. We love you so much, Mommy!*

Thanks to my husband, Kaustubh K. Marathe, DDS, for your love, kindness, and patience. Thank you to my four-month-old twin girls, Sonali and Manali, for sleeping almost all night, staying healthy and happy, and allowing your mommy to finish her book. I wrote this book during my maternity leave, sometimes trying to proofread while holding my twins tight. Often, I could only hold one baby at a time while I tried to type. Thanks to the girls for working with me.

Introduction

This book has been a very special endeavor for me. It has given me the opportunity to share my story, which is a journey from unimaginable horror, heartache, and disappointment to sweet victory. I am a survivor of the Killing Fields operated by the Khmer Rouge, also known as the Cambodian genocide. I am one of millions of inhabitants who in 1975 were forced at gunpoint to evacuate on foot into the countryside. Before you read on, I want to introduce myself. My name is Sopheap Ly (pronounced *so-peep*; the *h* is silent). My friends and colleagues call me "Sophie" because they tell me that pronouncing my real Cambodian name is like trying to say the alphabet backward.

I achieved a dream that at times almost seemed beyond my reach. I was once a child slave picking rice for fourteen hours a day, and now I am a medical doctor graduated from Howard University's College of Medicine in Washington DC. Through faith, a determination to never give up, and the memory my father's words—*Your dream is never beyond your reach*—my sweet dream came true. Not a day goes by without me reflecting on my graduation and the moment I walked across the stage and shook the hand of the dean of the medical school. Since then, I have continued with my work as a dedicated physician of internal medicine, committed to the health and well being of those I serve.

I am perhaps best introduced by the letter I wrote to the dean of Howard University's College of Medicine shortly before I graduated from medical school:

My name is Sopheap Ly. I came here to the United States as a refugee from Cambodia when I was 16 years old. My formal education started in the United States. I did not have formal education because as a child from the age of five to nine, I lived in the Killing Fields of Cambodia, picking rice fourteen hours a day. When I arrived in the United States, I knew very little English. I kept a dictionary in front of me while I watched TV—not because I wanted to be entertained but because I wanted to learn English. I worked several jobs to help my family pay the bills, but I never lost my focus. I believed the American Dream. I believed that America was the land of opportunity and if I worked hard and didn't give up I could become a medical doctor. No one could stop me. Nothing could stop me. People in my community told me that I was too poor to stay in school. That would not stop me. I would prove them wrong. I would stay in medical school, and I would graduate.

Now that my graduation day is about to arrive, I would like to thank Howard for believing in my potential, and giving me hope and the opportunity to pursue my dream of becoming a medical doctor. Not only did I learn medicine at this institution, but I also learned much more. I will never forget the lifelong friends I have made here. I am grateful for what the school has taught me—to be strong, to work hard, to work smart, to respect others, and especially to be humble. I am indebted to the students and faculty who encouraged me to work hard and to never give up.

Perhaps most importantly, I want to express my gratitude to hundreds of people who over the last 14 years in America have helped me pursue the American Dream. America is indeed a land of opportunity.

God Bless America.

1
Evacuating The Good Life

I was born in Phnom Penh, the capital of Cambodia. I do not remember much about my childhood in Cambodia before the Khmer Rouge took control of my country. I remember being happy because I knew my father loved me very much. I was the apple of his eye. I have memories of my father driving me around on his motorcycle to his older brother's house, which was a beautiful estate with lush farmland that had large tractors and shiny machines that harvested rice, corn, and lotus. He had forty people working for him. My favorite uncle was a wealthy landlord, and I affectionately called him Uncle Heng. His house was where I wanted to go on the weekends, because when I visited my uncle, he would tell me, "Take whatever is in my home. What's mine is yours." I didn't want anything from his home, but I did want lots of fresh lotus flowers from his garden. I remember carrying lots of fresh flowers back to my house and putting them in a vase. I will never forget how every year during the Cambodian New Year my uncle would stuff lots of large bills in a big red envelope and give it to me.

I loved my Uncle Heng for his generosity and kindness, and my father did too. I remember my father telling me, "Sopheap, this motorcycle you are riding on is a gift from Uncle Heng." My father was so proud that his brother had bought him such an expensive classic motorcycle for his graduation.

During our special visits to Uncle Heng's house, my father shared his dream with his brother. His dream was for little Sopheap to serve in the medical profession. My father looked to his brother for inspiration and guidance because Uncle Heng had raised him after their parents died at a young age. My father later told me that my uncle said, "Sopheap will pursue her dream, and she will succeed."

Like Uncle Heng, my father always encouraged me during my early years, sometimes whispering in my ear, "Never ever give up, no matter how terrible things might be. Always follow your dreams. They are never beyond your reach." His words would serve as a beacon of hope and light during my darkest nights.

Before those dark nights, my family life was filled with love, laughter, and fun. I was four years old when my parents enrolled me in a French private school in Phnom Penh. In Cambodia, a developing country, children usually did not go to school until a much later age. At four, I was quite young to attend school, and my mother sat in my classroom every day and wiped my tears, as I was very unhappy about being there. My dad, mom, sister, aunt, and I lived in a large red brick home in an affluent suburb of Phnom Penh. We had a nanny who was supervised by my aunt, who stayed with us while my parents went to work. On the weekends, we went on all-day shopping trips and visited relatives in the evening. One night, our cousins, aunts, and uncles gathered in front of my grandparent's house to watch the Chinese New Year dragon parade in the month of January. It was a joyful time and very exciting, as bright colorful lights—red, orange, yellow, green—flickered and flashed against the night sky, while a large dragon danced wildly in the street just a few feet from where I sat. I was being held tightly and securely in my father's arms as he sweetly kissed my forehead and cheeks while cuddling me in his lap. The fantasy of the celebration was beyond my expectation, and I couldn't imagine that life could ever bring anything but this kind of happiness and joy.

It was just a few months later that Khmer Rouge soldiers stormed our home late one night with guns and swords, demanding that we leave our home. I'll never forget that night. With only the clothes on our backs, my father, mother, sister, aunt, and I were ordered to leave our home by men with guns. After a long walk, we were herded onto a train like cattle. None of us had any idea that we would be taken to a distant farmland and nearly worked to death harvesting rice for four years before we would ever know freedom again.

The conditions on the train ride to our new destination were cramped and very crowded. As I looked around, people were compacted in like matchsticks in a small box. Without food, water, or bathroom privileges, some sat on the cold floor for hours while others tried lying down to sleep, hoping that when they awoke the nightmare would be over. The loud droning sound of the train wheels over the tracks screamed in my ears; crying babies and children wailed with hunger; women wept for lost loved ones, husbands and fathers stared blankly with anguished faces, as they could only anticipate the horror that awaited them and their families.

Everyone in our family, close and distant relatives alike, rode the train for a day and a night until we arrived at a remote farmland somewhere in the middle of very wet rice fields. We were pushed and shoved off the train like animals going to slaughter. Harsh speaking soldiers ordered us to get to work in the fields—if anyone had a problem, they would be shot or decapitated. We were now barefoot as we made our way to the rice fields under cruel and watchful eyes. It was then, at the age of five, that I entered the world of slave labor. I was told that I was to work at least fourteen hours a day, seven days a week, in the watery rice fields, while the adults were to work eighteen hours a day, seven days a week.

Dozens of families worked together to build crude shelters made of coconut and bamboo sticks or any tree that could be found that had leaves, sticks, and branches that could easily be pulled off. The shelters had dirt floors, and our beds were made out of bamboo sticks. We had no blankets, pillows, or any kind of warm covering.

Everyone slept in the same outfit they had on during the forced evacuation. New clothes were given only when the old clothes were in shreds or when the children's clothes no longer fit them. Black jumpsuits made out of thin cloth were handed out to us. I did not like wearing the jumpsuit—it was ugly and too big.

As I lay on my bamboo bed late at night, I wondered why this awful nightmare was happening to me, to our family, to everyone I loved and cared about. I thought about my favorite Uncle Heng and wondered what the soldiers had done to him and his family. I later found out that the Khmer Rouge had viciously separated my uncle from his family. His teenage son, Ratana, witnessed the handcuffing of his beloved father in front of his estate. His son watched angry soldiers cruelly tie his father with thick red ropes and then blindfold him with a heavy black cloth. Uncle Heng's face turned purplish black as he lost oxygen. Oxygen deprivation could cause death within several minutes even in a healthy strong adult. Angry soldiers hurriedly shoved my Uncle's limp body inside the four-wheel-drive truck. His teenage son helplessly watched the truck careen wildly on the unpaved road as it traveled about eighty miles an hour. Such speed and crazy driving was an attempt by the soldiers to confuse their victim, in the remote chance that he was still alive—and he was. But not for long, as the evil soldiers shoved his alive but greatly weakened body into a small rice sack and threw him into the Mekong River, where the blackbirds would later consume what was left of his flesh.

Such heartless violence against the innocent was difficult to comprehend. While trying to fall asleep at night, I thought over and over, "Why is this happening?" I could not go to my father for answers or comfort, as he was not in our family's shelter. The soldiers thought it was best for husbands and wives to be separated from one another, so my father was taken away from us shortly after we arrived. Now, I could only remember his tender touch, sweet kisses, and encouraging words: *Never ever give up your dream. It is always in your reach.*

It was difficult for me to fall asleep because I was afraid of the war situation we were suddenly in. So I spent a lot of my time looking through the crack in the wall made of bamboo leaves, watching the soldiers walk around outside our shelter with their long guns draped over their shoulders. The big guns hanging across on their bodies were very frightening. I didn't like their clothing—black pants and shirts, with small red-and-white checkered scarves tied around their necks, and dark caps on their heads. Seeing the toes of their black boots, I wondered who they might be kicking next. In my five-year-old mind, the soldiers were like gross aliens from another planet. Sometimes the aliens would pop their heads into our shelter in the middle of the night to check whether we were sleeping or plotting our escape. Anyone caught talking quietly would be dragged out of bed and killed for committing the "crime of speaking evil against the Khmer Rouge."

After a while, I could not cry any more tears. I had no more left. How strange and scary my life had become—

no more nanny, no more soft blankets or sheets, no more shopping trips, no more visits to grandparents' house to watch a colorful dragon parade, and no more father to hold me tightly like before and whisper in my ear, "Never give up on your dream." With only hope and faith to hold on to, I let my father's words burn like a bright light through my dark night.

2

Bloodstained Rice

Everyone working the rice fields was sleep deprived because we were only allowed to sleep a few hours each night before we were rudely awakened at 3:00 AM by soldiers harshly shouting, "WAKE UP!" All family members were jolted awake and rushed to get out of bed, knowing that if they were even a minute late showing up to work in the fields, they would be at risk of losing their life.

Work began at 4:00 AM, and most days it was not over until very dark. We were not allowed any breaks, and the only time we were allowed to eat was at the end of our day. Our food was a small portion of watery rice and a bowl of rotten vegetables; it was hardly enough to keep us from feeling hungry. I felt the most hunger during our early morning walks to the rice fields. There were many times when I would stumble over rocks, large roots of trees, or logs because I was so weak and because it was dark and I could not see the road very well. Even though there were times I scraped my knees and hurt myself, I couldn't stop to recover. I didn't want to be killed for arriving late.

Hunger and darkness were not the only things I struggled with on the early morning walk to work. Leeches were eager to feast on my flesh whenever I stumbled and fell into the watery rice fields. I remember landing with my face down in the water, barely conscious and unable to stand. After several minutes, I stood and discovered dozens of black leeches all over my body sticking their tiny tentacles into the flesh around my nose, ears, eyes, neck, arms, body, and legs. Relief came after I scraped the leeches off with a wooden stick that was about one inch in width. Only a strong stick would work, as the leeches were very stubborn and extremely difficult to remove.

The early morning walk routine was strictly kept for four years. We worked in the rice fields seven days a week, except when it rained heavily. The rain made it impossible to harvest rice, so it was on those days that we were allowed to kill and roast rats on our homemade spit. The rats were about the size of a small dog. I remember seeing them pop their heads out of large holes. The rats would come out and run right across my bare toes. Our family learned to outsmart them by setting traps. Once caught, we put them to good use. They became protein-rich food for our starving bodies.

When it was not raining and we were not roasting rats, I worked alongside other children in a different rice field from the adults. The adults were made to work eighteen hours a day, while the children worked fourteen. Every day, I picked tiny heads of rice under the scorching sun without any kind of protection. Many times, the heat was unbearable. One day, when I was about eight years old and working in the rice fields, I decided that I'd had enough. I was going to run away and find my mother who was far away in the adult rice fields. I had no idea where she was, but that didn't stop me. I did not care if the mean soldiers saw me—I was in too much pain, burned by the sun, sick, and tired of my life. I believed that running away would be the answer to my problems, even if it meant that I might be shot by the soldiers or get lost in the rice fields and die of hunger. All that mattered to me was getting away and finding my mother. For the next several hours, I walked through wet slushy rice fields often collapsing with exhaustion and landing face down in the mud, soaking my clothes with water and dirt. It was a miracle that I survived my brief escape and was not noticed by any of the soldiers. Many hours later, I found my mom. She was very shocked to see me suddenly appear, all red with sunburn, looking like an overripe tomato. She immediately told me to be quiet and quickly tried to hide me from her supervisors and the other soldiers. She gently and quietly convinced me to go back to my field with the other children because she feared that if I did not return, all our family would be shot and killed by the soldiers. I was not happy to go back, but I wanted to live and wanted my family to live too.

I made it safely back to the rice fields, and no soldier noticed that I had been missing for hours. The next day, I was forced to attend school. There was no classroom; class was held under shaded trees, and there were about twenty children who sat there in front of a black chalkboard, wondering why they had to endure a boring lecture about communism. I looked around and thought that working under the hot sun in the rice fields was better than hearing the teacher. The teacher must have sensed my thoughts; the next moment, I felt her hand on my head as she slapped and pushed me to the ground and placed her large hands around my throat. She gripped my throat for several minutes, and at one point I almost passed out. I could not breathe as I struggled for my life. When she loosened her grip, I found my voice and began to cry. Huge tears rolled down my cheeks, and I watched them fall on the dry ground. I could not stop crying. I was so frightened. I must have sobbed for close to half an hour, and just when I thought the teacher was finished with me, she tightened her grip around my throat again. Finally, she gave up and let me go back to my shelter alone. I never went back to the teacher's class again. I told my mother what happened, and she said, "Don't tell anyone because our family would get hurt or killed for talking about it." I was young, but I took the consequences of telling anyone about the incident very seriously. I have kept my mouth zipped very tightly until the writing of this book.

The next day, I gladly went to my work in the rice fields, though I could not forget the horror of what I had gone through. After I returned home from my work late that evening, I could not sleep. The mosquitoes were buzzing over my head and biting me. Red welts formed on my body, and I buried my head under my arm and prayed that the mosquitoes would stop their attacks. The attacks soon ended, and I fell asleep exhausted.

When we weren't battling mosquitoes at night, we battled the rain. Many times, it rained during the night and water poured in through our very thin leafy roof. I would wake up with my clothes soaked, and I still had to walk in the dark in the early morning to work in the rice fields. No extra set of warm clothes was ever provided to me or anyone else in our family.

During the four years, my family and I were prisoners forced to work night and day in the rice fields. Our suffering never seemed to end. I carried with me a great sense of loss and hopelessness and wondered if things would ever change or if I would ever achieve my dreams of a better life. Then I would remember my father's words: *Your dream is never beyond your reach.* I had hope, even though not a day went by when I didn't hear of someone who was taken away to be killed or who died of disease or starvation.

My grandparents were among those who died. I watched sadly as they lay in our shelter dying of starvation. I remember seeing my grandparents lying in the hut in extreme hunger. They chose not to eat the rats for food, as it was against their Buddhist religion. Without the extra food, they became very weak and did not talk or recognize their grandchildren anymore. Their eyes were sunken, their skin was so dry, and their gray hair continuously fell out. They had no more muscle that I could observe. Their bellies were flat like a pancake because they had no extra body tissue to spare anymore. It was so painful to watch. They had once lived a very happy life as grocery store owners, and they were generous and loving toward all—especially their family. In their last days, my grandparents could not lift their hands to wave good-bye to us as we went off to work in the rice fields. One day, my sister and I came home and saw my grandparents being carried on a thin metal sheet by my mom and aunt, who were already very weak with hunger themselves. We watched tearfully as my aunt and mom dug only a shallow grave for my grandparents; they were too tired to dig a deeper one. My sister and I sat on the broken steps of our bamboo bed feeling helpless, confused, and lost in our sadness.

Most of the time, after working all day in the children's rice field, I was left alone with my sister because my parents were forced to work from dawn to dusk far away from the house. We sat for hours in our small coconut hut with nothing to do. Sometimes we would dig holes or make up silly games to pass the time. Other times, we would just sit and stare, praying for our nightmare to end. Our parents would not come home until dark. It was our routine to sit and wait on the broken wooden steps of the bamboo bed in our hut until our parents came home. We were always hungry—for food and for love.

One day, my dad did not come home. We waited several days. When I realized that I would never see him again, I cried for days. Even after my father went missing, I kept a small hope that one day he would return. All I had left were memories of his love and his words, which played over and over in my mind: *Never give up on your dream. It is never beyond your reach.*

I long for my dad's soft touch. I long for his kisses and sweet words with so much hope. They are so meaningful to me.

I started crying non stop when I knew for sure that he was dead and would never return, never return, never return. I long for more of his love but there would be no more love--my dad was gone. He was killed. He was killed innocently. I remember his words vividly. My dad told me everyday that he loved me more than he could say.

I remember the last moment with my dad. It was a very late evening and the air was fresh with rain. He and I were sitting outside of our broken hut--rain water dropping on our happy faces.

We were together with the pure and raw love of father to daughter. My dad and I sat in the rain on a few branches because our hut had a broken leafy roof that leaked with rain water any way inside. Outside the rain soaked us.

My dad held me tight and kissed me and he told he loved me more than he could say. Those were the last few words he told me and the last moments we shared together.

I lived with my dad only seven years of my life. The duration was so short, sweet, and so unforgettable. I have so many happy memories of my dad. That seven years has had so much impact in my life.

Through our relationship I learned that it is not about the quantity of time but the quality of our good times together that has shaped me.

Months later, I learned that my father had been decapitated and cut into three pieces. I was so hurt and angry that the soldiers had acted so cruelly and wickedly in killing my father. He was a good man. Everyone who knew him loved and respected him. He was known for his integrity and his great love for his wife and children. He died speaking the truth about his profession. Under the Khmer Rouge regime, anyone who was educated or a professional was killed; only the weak-minded were allowed to live. Many Cambodians who experienced the Killing Fields are still alive, unlike my father, because they lied to the soldiers when asked about their work. My father refused to lie, and he died for it.

3

Free To Struggle Forward

On January 7, 1979, the light was beginning to dawn. Vietnamese troops began to occupy Phnom Penh, Khmer Rouge power was weakening, and our days of roasting rats were soon to end. No longer would we live in fear of being killed or spend eighteen hours a day harvesting rice. After a night of gunfire, our nightmare was over. I remember being very afraid, lying still on my bamboo mat, hearing the sounds of battle in the rice fields. I wondered whether I would wake to another day of life or this would be my last. I prayed and asked God for deliverance, and when I woke up the next morning, it was to the sound of Vietnamese soldiers telling our family that we could leave and return home.

However, we could not really go home because during the Khmer Rouge invasion our family home had been destroyed and our second home had been taken from us. After we were set free from the rice fields, we had to live on the bottom floor of a stranger's home, where we had no walls or door. Twenty-two members of my family—with ages ranging from infant to forty years old—lived for almost a year in that 1,200-square-foot dwelling. Though it was small and often inconvenient, it was better than the bamboo and coconut leaf hut we had lived in for almost four years under the Khmer Rouge.

After a year of working and saving money, our family was able to buy enough bamboo to build a long, one-room, rectangular house that fit our family. To separate from each other, we hung curtains to make different rooms. Our water to bathe and cook with was taken from a twenty-foot well that we dug. The water was dirty, so we had to strain it and then boil it in large buckets before we could drink it or cook with it. We made a bathroom from dried brown coconut leaves. It rained heavily in the Cambodian province of Siem Reap, so we had to reconstruct our bathroom every few months. To bathe our bodies, we had to go to the well, carry our water, and then we could wash after putting a chemical in the water to purify it.

We lived in our bamboo-and-coconut-leaf rectangle house for about a year. Despite our struggle for privacy, warmth, and cleanliness, we were thankful to be free. During the time I lived in our rectangle house, I went to work selling shoes, candies, and clothes in a swap meet market. My family bought these goods for a low price from Cambodian smugglers who purchased them from Thai people at the border. It was this business of buying low and selling high that enabled our family to earn money to rebuild our lives. I worked seven days a week, from 7:00 AM until 8:00 PM.

When I wasn't working, I went to school part time. I had to walk on a dirt road for two hours to get there. Even though my classroom was very hot and had no air conditioning, I was determined to get my education. The school walls were made from concrete, and the roof was made from zinc, which is a great conductor for heat. If anyone in the class showed weakness—crying or complaining—my teacher would bring out a broomstick and hit the student. Thankfully, this teacher never tried to strangle me or any other student like the teacher I had back in the Khmer Rouge rice fields. However, I did see several of my classmates cry with pain when they were hit with the broomstick. Sometimes they were hit so hard that they were left with bruises and welts.

Instead of becoming discouraged with my teacher or my school environment, I tried to make the best of the opportunity I had to learn. I was determined that I would never give up on my education and that my hard work would one day pay off. I held onto my father's words: *Your dream is never beyond your reach.*

With my father's words in my heart, I continued on with my learning in the concrete school for about a year, and it was not until I moved to the United States when I was sixteen years old that I had my first opportunity to have a formal education in a classroom without fear of being strangled or beaten.

After we had been living in our rectangular home for several months in Siem Reap, where I had begun attending school and was beginning to enjoy learning, our family had saved enough money to leave on a boat for Phnom Pehn. We wanted our own home and a better life.

The boat we rode to Phnom Pehn was large like a cruise ship without the luxuries or space. Thousands of Cambodians were crowded in like sardines in a tiny can. I stood on the deck staring at the Mekong River, watching the water turn over and over as our boat edged its way through the murkiness. It was this river that had become the watery grave of my dear Uncle Heng who had been horribly tossed into it as if he were a piece of foul-smelling garbage instead of a precious human being. The Mekong River had received countless victims of the Khmer Rouge's rage, and now it held the skeletal remains of thousands of bodies that at one time had floated to the surface only to be eaten by blackbirds eagerly awaiting to engorge themselves on that which would have never been their food, had evil not existed. I shook off my memories, naively hoping they would not surface again. I focused my mind on the large and small fish that joyfully leapt out of the river. Seeing such happy fish was the only thing on the boat ride that I found interesting and fun besides seeing my German shepherd dog, Black, run around like he owned the boat. Viewing sunsets from the boat was spectacular, and my hope was renewed that a better life awaited me somewhere beyond the Mekong River. For the first time, I felt that freedom and true happiness were not far off.

We were on the boat for two days and a night. When we arrived in Phnom Pehn, my family told me that we would soon bc leaving our beloved home country and moving to a new one, where a better and brighter future was possible. My family feared that Cambodia's peace would not last for very long before another evil uprising might occur.

4
Home to a Different Town

My family and I found a small apartment in Phnom Pehn. The beautiful, spacious home that we lived in before the forced evacuation had been destroyed, and now we had to live in a cramped, rundown apartment with lead paint on the cracked walls. My family purchased the apartment for very little money, but we had to pay the high price of living in an 800-square-foot space with twenty-two of us. There were no bedrooms in the apartment, just one long hall. The apartment had no electricity, a broken down toilet that barely flushed, and running water, but no gas or electric stove. We used wood to cook, and there was no ventilation system, so smoke circulated everywhere. Sometimes the smoke was so bad that we could not breathe and had to hurry outside to get a breath of fresh air.

With twenty-two of us crammed together in a long hall, sleeping was very challenging. Many of us had to sleep on the floor because there were not enough beds. There was no privacy, and we could never completely escape each other's snoring or other noises. Few of us ever got any sound sleep.

Our apartment was also noisy with sounds coming from outside on the street in the heart of the capital. Cars and motorcycles zoomed by at all hours of the day and night. When the street calmed down, the mosquitoes began their incessant buzzing, which often would keep me up for hours. As I lay on my bed, unable to sleep, I would look up at the ceiling and often see a lizard stuck up there. It was a strange creature that made loud "cricking" noises. I wondered if it was carrying some deadly disease. Despite the lizard and the unpleasant surroundings, I was very thankful to be out of the Killing Fields and to be given a second chance at life.

Another unpleasant condition our family had to overcome was our water supply. It was extremely limited by the government. Running water from the faucet was only released at certain hours of the day, so whenever it was—and we never knew exactly when—we had to scramble to get our supply of water to meet the needs of twenty-two people for cooking and bathing. Unfortunately, we only had a small number of buckets to store the water. We were forced to take very short baths, sometimes only once a week, with very little water. Flushing the toilet rarely happened, as that took too much water. Hearing snoring at night was a minor problem compared to the stench of human waste. We later found places outside to relieve ourselves or used the toilets of our neighbors who were able to flush.

Not only was our apartment in severe need of repair, but the whole capital city of Phnom Penh was filthy and in disarray. After the evacuation, during the four years of communist occupation, the city became a ghost town. All businesses, schools, and neighborhoods were shut down. Now the city was struggling to come back to life, businesses were slowly returning, and schools were beginning to open. I was pleased that soon after we moved into our apartment, I was able to attend a school a short distance from where we lived. Every morning, I walked eagerly to school with my sister, anxious to get there because I could not wait to get back to learning. After school, however, I could not study. I helped my family sell clothes and medicine in a swap meet market. Whatever amount of time I spent in the classroom, I had to learn quickly, since I could not study after school and did not know if there was a library. When I was not in the market selling goods, I cooked food for my family. As a nine year old, I was also responsible for hand washing my own clothes, even if that meant I had to walk a distance to the river. Sometimes my younger sister came with me to do my washing, as it was my responsibility to take care of her, too. Many times, I struggled with envy watching other children who seemed to have more carefree lives. Unlike me, after school they were able to study or play with dolls and other children in the neighborhood. I couldn't do those things because my family was too poor. Even though I was poor and did not like that I had so many chores, I was blessed to realize how valuable and important an education was. At an early age, my father had instilled in me a love of learning as well as a desire to pursue my dreams. "Education," he told me, "will enable you to better your life and leave poverty behind."

My relative, though well meaning, pushed me to learn too fast. She bribed one of her friends who worked at a public school to put me in a higher-level class so that I would get my education faster. But the truth was, I did not belong in that class. I was only eleven years old and already very small for my age, and when I looked around, I saw my female classmates with large breasts. I felt embarrassed and wanted to go home. I also was not prepared to do such advanced level work. I did not have the early educational foundation to be able to master the math and science. I struggled so much in those few months, and I cried at home when I was given homework that I could not do. My relative tried to help, but she did not understand or have sympathy for my frustrations. An uncle stepped in to help me, trying to act as a father figure, but he also did not have much compassion for my struggle. My uncle was very impatient, and several times he beat me up while screaming curses and calling me terrible names. I quietly cried myself to sleep after my homework "help" with my uncle. I prayed and dreamed that I would be able to escape to a Thailand refugee camp where I could find the right kind of help to learn basic math and science.

5
Seeking Refuge in Thailand

My family had the goal of leaving our apartment and our country from the day we first arrived in Phnom Pehn. They had seen pictures of America from distant relatives and heard about "the good life." All of us wanted a share in the better life, but especially me. It was my dream. My father's words—*a dream is never beyond your reach*—continued to live in me day and night, always giving me much hope and encouragement to keep moving forward.

A guide was hired to smuggle our family out of our apartment and take us to a province near the Thai border. He came for us late at night and took us to a small hut in the middle of a rice farm, where we waited until the sunset. Our family huddled together, and everyone did their best to try to sleep, but I could not sleep. I was too scared of what was ahead. Thoughts raced through my mind about what might happen to me on my journey to another country: *Would I survive? Would I make it to America? Would I be successful? Would my dream turn out to be beyond my reach?* I was told to hide in the hut and not to make any noise because if any of the soldiers from the new regime that had taken over Cambodia came by and found out that our family was planning to escape, we would likely be killed immediately.

There was a silent moment, so sullen in my life, as the guide told me to get on the small boat in the flooded rice field. I was very afraid to get on board, since I did not know how to swim. But I kept thinking about my father's dream for a better future, and I knew I had to leave my beloved country for a good life. Looking at the horizon, I saw a gorgeous sunset with so many colors in the sky. It touched my heart how beautiful Cambodia was; yet I could not stay there. I thought about my father's death and knew I was leaving his remains somewhere unknown in Cambodia. I was very sad to leave, but I strongly believed that my father's spirit, powerful dreams, and hopeful words would never fail to be with me as long as I was alive. I knew that I could achieve my dream and dared to get on board, headed west. We got off the boat and walked for many hours on a dark road, with our guide in the lead. So began our all-night journey. It was very dark because the moon was not out, so we had very little light to guide us, except for our small flashlights. We were careful to be very quiet, as we did not want to wake up any soldiers on the way. Morning came, and we were exhausted from walking in the forest all night. Yet we were excited to see a beautiful tropical rain forest surrounding us with tall trees, large and small lush green leaves; roots from trees twisting around other trees; different insects busy moving in various directions; grounds thick with very green vegetation; and birds sweetly singing. This incredible scene reminded me of how breathtaking the Cambodian and Thai rainforests were and how blessed I was to have just a few moments of respite.

However, the beauty was soon jarred when Cambodian soldiers suddenly jumped out of the bush and pointed long rifles at our faces. They raged and screamed at us, demanding that we give them our belongings—gold, diamonds, and money were what interested them the most. No one in our group said anything. We all pretended that we were deaf and had not heard anything they said. They angrily shouted, "You all are heading west?" Again, we said nothing. The soldiers finally left as our "deaf and dumb" trick had worked. We were thankful that no harm was done to any of us in our group. I later wondered if our smuggler had bribed the soldiers with gold or something else so that we could continue on our journey. But whatever caused the soldiers to leave, I was thankful for it.

We continued to move forward through the rain forest, hoping that we would not run into any more soldiers. Our nerves were frayed, our bodies exhausted, and our clothes wet, but we pushed forward, determined to never give up. I was perhaps the most determined, as my father's words continued to give me strength and courage: *Never give up on your dream. It is never beyond your reach.*

As we finished the last leg of our journey to the Thai border and refugee camp, I found myself stumbling over large logs and running into leeches as I had many times before when I fell into the watery rice fields. But I refused to give up. I wanted to reach my dream.

We arrived in the first refugee camp, Chum Rum Tmei, in the middle of the day. We were told that it was only a transition camp and that we would soon have to leave. I didn't care, as long as I was one step closer to reaching my dream. I was happy to live there for one month.

Soon after we arrived, we were asked to register our names with the UNHCR to get a supply of food and water. Our guide led us to a hut and bed made of bamboo leaves, which I soon discovered had a five-foot python coiled underneath it. By this time, very little could scare me, even a five-foot snake. I ignored the python, and soon our unwanted visitor left.

We stayed in our hut for one month. Living conditions were very crowded, and there was no electricity, running water, or toilet inside the hut. The toilets were outside, and flies and maggots were crawling in there. All of those who had struggled to come to the camps were now at great risk of disease and death, as the living conditions were extremely unsanitary. Sadly, many in the camp died of infection before their dream of coming to America was realized. Thankfully, no one in my family died.

Clean drinking water was delivered in a big truck by workers from UNHCR (the Office of United Nations High Commissioner for Refugees). The truck with water only came once a day, early in the morning, and the water was not enough. Most of us went days without a shower. Since water was so scarce, we had to dig our own well and hope that we would find water. When we did, we had to boil it so we wouldn't get sick from drinking it.

6

Another Camp Closer

A family member had paid a Cambodian expert guide to smuggle us out of the country. Now we had to look for another guide who could take us to our second camp. We had to move on to the next camp in Thailand, called Khao-I-Dang camp, so that we would have a better chance of being accepted into the United States. This time, I was not afraid to escape. I had already made it through one hell, and I figured I could make it through another. I also began to believe that there had to be a reason and purpose for what I was going through. Remembering my father's love and encouragement kept me going at times when my hope for a better life seemed lost. I took courage as I kept his words close to my heart. I pressed on, willing to take whatever risks were necessary to achieve my dream of moving to America.

The trip to Khao-I-Dang refugee camp was just as dangerous as our earlier escape to the first camp. We left in the middle of the night, but this time we were not trudging through a rain forest. We trekked through red dirt, dry shrubs, and an occasional small forest of trees, as we kept in mind the risk that we could be captured at any moment by Thai soldiers who were always on the lookout for pretty young teenage girls to rape. If the girl did not cooperate, she would be at serious risk of losing her life. Our guide told us that the soldiers kept a lookout for their next victim by hiding in the treetops around us and flashing their lights on our path.

After walking for over eight hours, our guide brought us safely to our destination. We entered the refugee camp in the very early morning hours. Our guide carefully cut a small hole in the metal wire fence that surrounded the camp. He took out a small pair of scissors and began to snip away at the wire to make a hole that was just large enough for our group to squeeze through. Like mice trying to sneak past the hungry cat, we quietly passed the sleeping guards who were on a mission from the Thai government to seize anyone who was an incoming refugee to the camp. We later learned that it was common practice for a guide to bribe a guard should they awaken and see any unwelcome visitor. Most guides were prepared to bribe the guard to keep them from sounding the alarm. The only bribe that worked was giving the guards gold. After successfully making our way past the guards, our guide hurriedly rushed us into a bamboo leaf hut shared by another escaping family.

We could not openly carry any gold or any belongings with us when we first left Cambodia, but we knew the importance of having gold with us: It could save your life when used at the right time. On that trip, we hid our gold in the seams of our clothes. But this time, we hid our gold inside our bodies by carefully inserting like suppositories two-inch-long gold bullets that were one-quarter-inch in width. Before we left for the second refugee camp, I had placed a few of the gold capsules in my rectum. But somehow, by morning, I only had one capsule remaining, and my last gold suppository slipped out. So several ounces of my 24-karat gold had mysteriously dropped out, and they were not evident in my stool that morning after my bowl movement.

Later that morning, the family we stayed with in the hut took us to the UNHCR office, where we applied for food and for settlement in the United States, along with thousands of other Cambodian refugees. We would stay at the camp for three years until we were accepted to live and settle in the United States. Life was not easy in the camp, which had approximately twenty thousand refugees crowded into only a few square miles.

Not only were conditions crowded in the camp, but frequent violence was also a threat. Many times, Thai robbers with big guns rushed in to our area with strong intent to rob, rape, or terrorize anyone that stood in their way. Thankfully, no harm was ever done to my family or me. We protected ourselves from the robbers by digging a shallow well about five feet by five feet. It was our hiding place, and we would stay there all night huddled together.

We found it very difficult to breathe because of our cramped and suffocating quarters.

To say that our three years in the refugee camp was unpleasant would be a severe understatement. There were at least a couple of days a week when we did not receive water or food because the delivery truck failed to show up, so on those days we went to bed feeling raw with hunger. Food and water were rationed daily. We rarely had enough to eat, and our bodies became weaker and highly susceptible to infectious diseases. Being exposed to heavy dust, dirt, and very dry weather did not heal and strengthen our bodies. One time, my thirteen-year-old body was very weak and tired. I developed a terrible infection with coughing, sneezing, sore throat, difficulty swallowing, and frequent fever. After suffering with my condition for a long time, I wondered if I would ever get better and achieve my dream of going to school like a normal, healthy child. Every day, I prayed for my health to be restored, and finally my prayer was answered. I was allowed to visit a health clinic supervised by several Thai doctors and nurses who prescribed me with different antibiotics, but none worked. I was then sent to a Thai ENT doctor, and she immediately diagnosed me as having chronic tonsillitis. Without much delay, I was taken by car to another province so the ENT doctor could perform the surgery to remove my tonsils. After I arrived in the modern hospital, I was not given a bed; I had to lie down on the cold concrete floor. I had no coverings because I was a refugee and could not afford better health care. I was put in a separate ward with all the other nonpaying patients, while the other side of the ward was filled with paying patients who rested comfortably in nice rooms and warm beds. I was very upset, hurt, and disappointed over the inferior medical treatment given to those who, like me, could not afford to pay for better health care.

Despite the conditions I encountered at the hospital, I met an angel: an extremely kind nurse who understood my pain and suffering in the camp and in the hospital. She paid extra attention to me and tried to do whatever she could to make me feel loved and cared for—bringing me water, food, and pain medication. She often made eye contact with me, and her smile brightened my day and gave me much relief and courage, despite my fears of going to the operating room. My being poor and forgotten did not matter to this nurse. We later became good friends, and she would smuggle toys and yarn projects, including "how to crochet" books and magazines into the clinic where I went for additional follow-up after my surgery. She knew I would welcome something to do that could keep my mind busy, active, and free from boredom. She believed in me and knew that I was a survivor and would achieve my dream of moving to America. Though I did not tell the nurse about my father's dream for me to work in the medical field, I think she sensed I was a kindred spirit. She became a positive role model and gave me strong encouragement to never give up on my dream, even if it seemed almost beyond my reach.

I was wheeled into surgery the day after I arrived at the hospital. I was fully awake in a sitting position during the whole surgery. I remember hearing the clipping noise coming from the scissors that snipped the infected tonsils out of my throat. After the surgery, I was wheeled back to the cold concrete floor where I lay down in great pain as blood dripped slowly from both corners of my mouth. I could not eat, but I did not remember if IV fluids were given to me. I was too tired and dehydrated to fully comprehend what was going on around me. The kind nurse visited me, tried to comfort me, and made me feel better. But she was at risk of losing her job if she gave me too much time or care because she was not assigned as my nurse.

The next day, I was discharged and sent back to the camp. I traveled in a small car and sat alone in the backseat for the whole trip. I was not offered anything to drink, and when I finally arrived back at the camp, I was eager to drink something cold. But since there was no electricity to run a refrigerator, I could not find anything cold or soothing for my aching throat. In my pain and desperation, I ate whatever I could find, even hard foods which caused my surgical site to bleed again. It truly was a miracle that I didn't die from bleeding and that I was able to resume my normal activities within a few days, which included going back to school.

I attended school and studied basic math, Cambodian language, art, and English—which I was most eager to learn because I was going to America, despite the harsh conditions all around me.

The camp was made up of thousands of bamboo huts that were placed very close together, and each hut was made from bamboo sticks with roofs made from coconut leaves. Hot days were numerous, and there was no relief from the heat as there were very few trees in the camp. On some days, the horrible, heavy monsoon rains collapsed our leaf roofs. When that happened, we went without shelter over us until we could rebuild our hut on a dry day. During the wet days, our clothes and bodies were soaked with water, and it was fortunate that we did not catch pneumonia and die.

Sanitation was another issue we had to face in the camp. The restrooms were not like those in other developing countries. We had to dig a hole to relieve ourselves, and nothing was ever done to dispose of the human waste, so flies and maggots circulated everywhere carrying all kinds of diseases. UNHCR staff tried their best to help and protect Cambodian refugees, but there were so many refugees coming to the camp that they had a very difficult time managing the waste. Large trucks were sent in to extract the waste and remove it to a landfill somewhere outside the camp.

Even though life was extremely difficult in the camp, I refused to give up my dream. There was not a day when my father's words did not speak to me: *Your dream is never beyond your reach.* I decided to keep going, despite the possibility that death might win in the end.

7
Approaching the Dream

After a series of interviews, our names were finally chosen to come to America. Through much trial, travail, and heartache, our time had finally arrived and my dream was going to be realized—America at last. We were being given a second chance at life, a chance to live the American Dream, which I would soon come to find out was far from a golden fantasy.

After our approval, we lived in a settlement located in the Bataan province in the Philippines. It was there that life began to feel safe and happy again, almost like it felt during the first five years of my life. at my family's home back in Phnom Pehn. I started to learn English and the American way of life. I liked our life in this third camp because we were free to walk around with no a metal fence, no soldiers with guns hanging on top of trees monitoring our every movement, and no robbers breaking into our camp to rape or steal from us. At this camp, we could leave at any time to go to the market; our relatives could send us money for extra living expenses; and we were allowed to make friends with the local Filipinos. Our lives almost felt normal.

During our six- to eight-month stay in the Bataan refugee camp, our only concern was earthquakes. At night, we were often jolted awake by earthquakes—some small, others big. Though many of the earthquakes felt very strong, they never caused any significant injury to anyone in the camp. What I enjoyed the most during my stay in the Philippines were the bright blue skies; warm, sunny days; and crystal blue waters that were edged by crisp white sand. The beauty of it took my breath away and stirred in me great thankfulness that life had finally become better for my family and me.

Soon my dream of being in America became reality. I'll never forget walking off the plane in Dallas, Texas, one very hot afternoon and seeing America for the very first time. The airport bustled with people. The office buildings were tall, many made of glass, radiating shine and sparkle. I tingled with excitement. As an idealistic and enthusiastic sixteen year old, I had very high expectations for a golden America where everyone worked hard to achieve their dreams for a better life and where people truly cared for one another. After all, I thought, the struggle for survival was not as great in America as in my home country. Surely, people in America try to make life the best they can for themselves and each other. I was soon to find out differently.

We were picked up at the airport by a relative. The route home took us by many homeless people sitting on the curb looking hopeless and lost. I was surprised and thought: *Why should this be? This is America.* I felt sad for those people. I wanted to help, but I was powerless to do anything. I learned right way that in America I would have to work very hard to achieve my dreams, lest I too fall in a slump and find myself living curbside. I had already come this far, and I would do my best to let nothing stop me, even sad images of Americans who had lost hope. The homeless people served as a powerful reminder for me. I didn't want to end up living in poverty in a country full of opportunity and hope like America. I would do whatever it took to complete my higher education and help people better their lives and improve their health. I made my decision then that I would become a medical doctor. My father's dream of seeing me work in the medical profession had now become very specific in my mind. That dream was within my reach, and I would not stop until I achieved it. Even though I did not have a large suitcase packed with clothes, money, or material possessions, I had a strong desire to fulfill my dream.

My relative allowed my family and me to move in to her apartment for a short time. School was not going to be an issue, since we had arrived during the summer. My time was spent learning about American culture and life by spending hours watching television. I was eager to learn English. I kept a Cambodian/English dictionary in my hands while I was watching TV and found myself looking up many new words. My vocabulary grew quickly as I was determined to succeed even if educational television was my teacher.

Fascinated by international and national politics, I listened for hours every day to *The Voice of America* on the radio, as was my habit when I was living in the refugee camps. I was anxious to find out what happened to my country, why it happened, and what was continuing to occur. As much as I wanted to know, I still struggled with the answers that were given. The things I heard as explanations did not satisfy me. In my mind, the root of ongoing conflict had to do with greed and power struggles coming from those who carry the biggest guns.

Other television shows, such as sitcoms and MTV, never captured my interest. I was too interested in achieving my dream, improving my mind. I did not want to waste one moment on anything that would not encourage my mental development. One of the fears I had growing up in Cambodia, and when I was living in the refugee camps, was that I would not be able to successfully complete my higher education and that I would be forever doomed to a life of poverty.

During my brief one-month stay in Dallas, I met many teenagers close to my age who spoke beautiful English. I wanted to be their friends, as I hoped their strong language skills would rub off on me. Because I was a very shy, I could not find the courage to approach them and begin a friendship. I was also scared of approaching them because many of the teenagers, some of whom were not native born, put on airs, as if they were better or more important than me just because I had come later to America than they had.

I visited a Cambodian family who were friends of my relative. This family lived in Houston. They were blessed with a nice home and had been able to achieve a comfortable life after many years of living in America. This hardworking family told me that they worked in maintenance and janitorial services at a major airport. For a brief moment, I thought that might be a direction I should go. It was possible for me to go to work immediately, make steady money, save it, and later buy a house. But then I remembered my goal of completing my higher education and becoming a medical doctor, my desire being fueled by my father's dream for me to work in the health profession. This was the best investment of my time, and the rewards of making money and buying a home would have to come later. I wanted to do work that would be meaningful to me, so I learned the value of delayed gratification.

Shortly after our time with the Cambodian family, we boarded a plane to California. We were headed to the Golden State.

8
Hope and High School in Santa Ana

We arrived at Los Angeles International Airport before noon. Another relative met us at the airport and allowed us stay in their attractive and well kept home in Fountain Valley, California. We stayed with them for about two weeks. My relatives applied for welfare for us, so we were able to move into our own apartment in a very low income and violent neighborhood in Santa Ana. Many gang members lived there. We could hear gunshots day and night, and loud banging hit our walls and doors. At times, I thought we were living in a war zone. It again reminded of my days as a child under the Khmer Rouge. I would wake up and find bullet holes in some neighbor's doors and windows, surrounded by blood stains, as well as broken windows shattered by gun blasts and green yard areas sprayed with blood stains. If the horrific and terrifying life-threatening sounds from the gang activity outside my apartment were not enough, there was also a railroad nearby. The train tracks were only a few feet from our bedroom. The roaring from the trains created additional noise throughout the day and night, so studying and sleeping were almost impossible. I only fell asleep because I knew that I needed rest to excel in my studies at school.

After school, I worked as a seamstress a few hours a day for about a year in a nearby shop. I was making about $3.75 per hour, which I felt was a lot of money, but it was still not enough to buy milk and other foods. Again, we went to bed hungry at night. I could not believe that I was going to bed hungry in America. America was supposed to be a land filled with dreams of paradise and prosperity, but I was again fearful for my life and having to sleep hearing gunshots blasting through the air. Though I was no longer in the Killing Fields, I was now experiencing a different kind of challenge: surviving in America as a poor person living in a bad neighborhood and lying awake at night wondering if an errant gunshot would kill me. Though I had my fears, I did not let go of the dream planted by my father. I was never going to give up my hope of becoming a medical doctor and making a difference in people's lives for good. I came to understand what great power lies behind a father's encouragement to his child: it is the gift of hope that they can achieve their dreams.

In addition to the memory of my father, I had a relative who was also a source of encouragement. She often told me that when she first arrived in America, she woke up at 3:00 AM every day of the week to go to work in a doughnut shop. After working many years, she was able to buy her own doughnut shop. Every morning on her way to work, angry dogs barked as she walked by them. She felt very afraid, but she did not give up. She was too determined to make a better life for herself and her family and to one day become a millionaire. Her perseverance and ambition became a real source of inspiration during my trials and tribulations. From her, I learned the value of sticking to my goal and working hard to achieve it. After several years, her wealth is skyrocketing and she still continues to work very hard.

Despite my challenging circumstances, I held fast to the American dream of working hard to achieve success. I believed that American law enforcement would uphold justice and protect the people. With this confidence, I was able to give my full attention to my studies at school and as well as my homework after school.

At Santa Ana High School, I was eager to learn whatever I could from my teachers and spent hours every night on homework. I was not afraid to ask for help when I needed it. This was my education, and I wanted to get the most from it. I was surprised and disappointed to see that most students did not share my passion for learning and growing. I eventually came to understand that they just didn't care about education in the same way I did. They wanted to take the easy road, and I was willing to walk the narrow path and believed with all my heart that it is not how you start out in life but how you finish.

I remember one female student who jokingly said in front of all of the students in my history class, "Is Sophie going to graduate with us?" It was like an arrow to my heart, even though I knew I would graduate because I was an A student and very determined to not fail any of my classes. I never did, and that was because I was blessed with teachers who inspired me.

My Spanish teacher, Ms. Daremple inspired me to love learning, especially studying the Spanish language. She told me, "Learn as much as you can so you can travel without having to rely on translators." I loved Ms. Daremple's enthusiasm and passion. I was one of her favorite students, and she gave me the nickname "Sponge." She told me that I soaked up everything she taught and that this was a very valuable trait that I should never lose. I never have.

Another teacher who inspired me was Ms. Saxton, my English teacher. She always reinforced in me the value of a higher education. She asked our class what we wanted to do after college graduation. It was interesting to hear the responses. One of my classmates said, "Ms Saxton, I want to be an assistant to someone." Ms. Saxton interrupted abruptly, "Why do you want to be an assistant to someone when you can be the boss?" Her words were forever etched in my mind.

Ms. Gomez was a most inspiring ESL teacher because not only was she very kind but she was also beautiful, intelligent, stylish and classy. Every day, she wore attractive suits and arranged her hair in cascading curls. Her appearance was as appealing as her instruction. She made us write in our logs every day, with the goal of helping us express our thoughts and feelings so we would become better students and achieve mastery of the English language.

I remember shaking her hand as she gave me several student achievement awards for academic excellence.

One of the most significant people who shaped my life greatly in the area of mental discipline was Mrs. Gordon. She taught me math for two years. Every day, she had her students writing formulas, mathematical laws, and theories more than twenty times on a paper. I understood everything that I was writing down, and I eventually received high honors in algebra and geometry on the Golden State Examination. She knew how hard I worked and saw that I could answer most, if not all, the questions in class, even on material that had not yet been taught. One day, she came to my desk after I finished my math homework one week early and asked me in front of all the students, "How many hours do you sleep every night?" I pretended that I didn't hear what she asked. I put my head down and continued to work on math homework. But she would not give up on getting an answer to her question. She asked again, this time a bit louder. "Sophie, how many hours do you sleep every night?" I finally answered, saying in a very low, soft voice, "Three to four hours. But on weekends, I sleep more hours." I could tell that Ms. Gordon was proud of me for my dedication and hard work, but I was embarrassed that everyone knew about my sleep and work habits and thought I might get beat up by one of my classmates for being too competitive.

I was determined to always put my studies first, but there was this one guy from Honduras who noticed me in ESL class. He came over to my table and called me "Lee," thinking that my last name was my first name. And then he couldn't pronounce my first name, Sopheap. I was flattered that such a nice, good-looking guy with a sweet disposition and warm brown complexion noticed me. One of the very first questions he asked me was, "What do you do for fun?" I felt very embarrassed because I was not having a whole lot of fun in my life. I stumbled with my words and said something like, "Well, I have been studying. I am always busy studying." He said in a calm, caring voice, "I always see you sitting on your chair for hours studying, and I have a basketball playoff coming up soon." I looked away and didn't know what to do with his interest; I went back to my books and never spoke with him again. I later regretted not allowing myself time for fun and relaxation. I had made work and studying too important. But I would come to understand the value of living in more a balanced way, where work and play are both part of a full, happy life.

At that time, however, I could not think about guys because I did not even have enough time to sleep. I was too busy working after school. Walking to work after school was exhausting by itself; plus I worked four extra hours as a medical assistant for two years while I was in high school and two more years in college in the same clinic, helping about thirty patients on my afternoon shift.

Photos

Author's note:

I do not have any pictures of myself when I was a baby. In 1975, when we were forced to evacuate our home in Phnom Penh, Cambodia, we were not allowed to bring pictures or any other belongings with us on the train to the rice fields. All the pictures left behind were destroyed by the Khmer Rouge, along with everything else. The youngest pictures of me are from my years in the refugee camps, where we wore used clothes given to us by UNHCR (Office of the United Nations High Commissioner for Refugees); I was very happy to wear the used clothing.

Standing on the far right, I'm walking to school with four other girls in the refugee camp in Thailand.

That's me on the far left, posing with classmates outside the classroom.

In the second row, I'm second from the left, sitting with my hair in a ponytail in our bamboo-crafted classroom.

The first three adults, *far left*, are our teachers, and the rest are my classmates.

My sister, Sopha, *right*, and I venture into the woods near the refugee camp in the Bataan province of the Philippines.

Trying to escape the heat, I play in a stream in the Bataan province refugee camp in the Philippines.

Found it! Getting a book off the shelf in the library at Santa Ana High School.

Even while relaxing on the grass at Santa Ana High School, I was still doing schoolwork—a total nerd.

My love for chemistry lecture and lab began in high school.
That's me smiling with my goggles on my forehead.

Sitting under the welcome shade of a big tree, I look up from my homework briefly to take a laughter break.

Smelling the roses while touring the University of Southern California with classmates Sylvina, *left*, and Ruth, *right*.

On the far left, I'm all dressed up for a special evening event for the GATE (Gifted And Talented Education) program at Santa Ana High School.

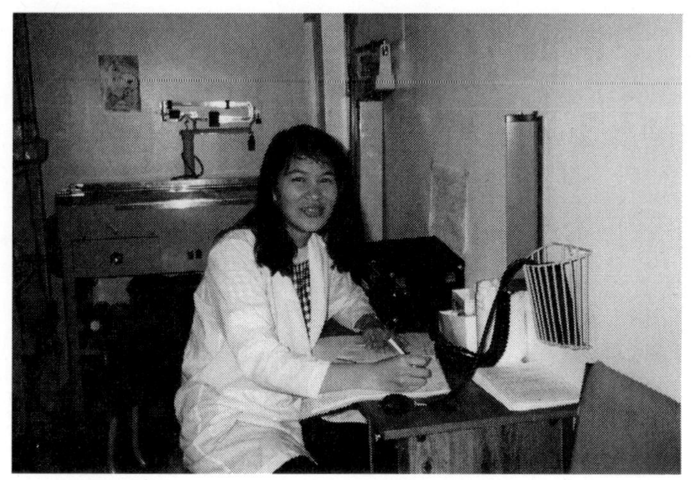

In high school, I worked as a medical assistant after school, earning about $3.75 an hour.

Taking a quick snack break with a coworker, *right*, at the medical clinic where I worked in high school.

A welcome photo op with the Close Up tour group on our trip to Washington DC during high school.

My high school graduation photo.
Someday, my twin girls will say, "Mom, that was you?"

The love affair with the chemistry lab that started in high school continued after I arrived at California State University, Fullerton.

Couldn't miss a chance to take a picture with my favorite chemistry professor at California State University, Fullerton, Dr. Gene Hiegel, *far left*.

My McNair/Star research group at University of California, Irvine.

Everyone look this way!
Snapshot of my Berkeley Undergraduate Research Institute group.

Attending the Student Health Profession Association banquet with fellow members (I'm on the far right).

On stage receiving one of several scholarship awards at California State University, Fullerton.

Berkeley Undergraduate Research Institute

We are preparing to take off from Banning, CA, on this small plane to Mexico to do medical volunteer work.

Proud to be an official U.S. citizen, I pose for posterity in front of a sign pointing the way to "naturalization."

My sister, Sopha, *right*, and I at Huntington Library in San Marino, California.

Graduating from California State University, Fullerton, with a bachelor's degree in biology and minor in chemistry.

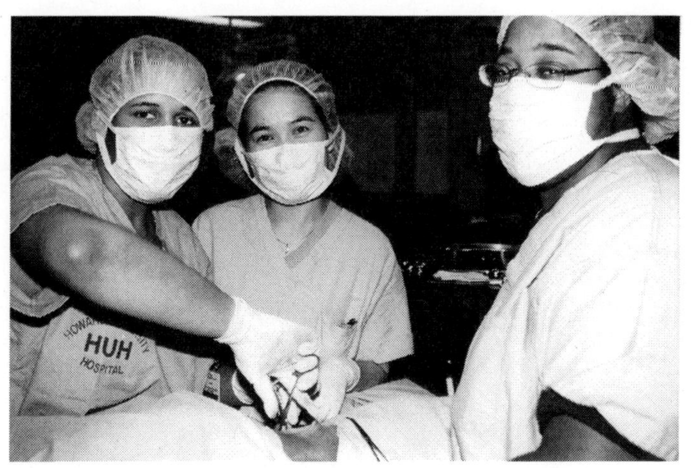

Howard University medical school classmates and I learn how to cut and suture pig leg skin.

Our resident teacher, *far left*, decides to take a picture with us after many hours of suturing pig leg skin.

My internal medicine rotation group.

Standing with classmates near anatomy lab.

Dinner with fellow medical students after a long, difficult exam brings smiles.

A group of friends from Howard University gather at our classmate Tiffany's apartment for a get together.

Aretha Makia, *left,* Kansky Delisma, *center,* and I in front of the Howard University Health Sciences Library, where we spent many hours studying together.

A shot of me with Howard University anatomy professor Dr. Mohammed Aziz.

Studying for my medical board exams at a Kaplan educational center in Washington DC.

My friend Dr. Mijeung Gwak comes to visit me from New York while I'm studying for board exams.

Love is in the air as I attend a friend's wedding in Maryland.

Wearing a dashing tuxedo at the wedding, my friend Aretha's 8-month-old son, Ebot, gets a kiss from Auntie Sophie, who loves him very much.

On our graduation day, Denise, *center*, Aretha, *right*, and I are happy beyond words.

In our last school picture together, Aretha and I are all smiles after graduating.

Walking across the stage to shake hands with the dean of Howard University's College of Medicine, my dream has come true.

Early birds. World renowned liver specialist Dr. Telfer B. Reynolds and I arrive in the cafeteria at LAC-USC Medical Center in time for breakfast.

9
The Golden State of Opportunity

After high school, I was very excited to attend California State University, Fullerton (CSUF). In the acceptance letter, the university stated that they wished for me to graduate from college with honors. But I was discouraged because I did not have enough money to pay the tuition of only $550 a semester. I spent hours in the library and the dean's office searching for scholarships. I wrote essays seeking scholarships of all kinds. My efforts finally paid off, and I received a Cal grant, a federal scholarship, work study grants, as well as private and school scholarships that almost covered my tuition and living expenses. I was thankful, and being granted the scholarships boosted my confidence that others in the academic community believed in my potential.

During my freshman year at CSUF, I took introductory biology—and I did not do well. I barely passed because I was too busy working many hours after school to pay for my college costs that weren't covered by scholarships. There were many semesters when I worried when my scholarship money would show up in the mail so I could pay my bills. Sometimes my checks would not come, and I would be on the phone calling to ask when it would arrive. I had a recurring nightmare that I was kicked out of school and could not register because I could not afford tuition. Sadly, my nightmare almost came true many times during my years as a student at CSUF. One time, I had a dream that my father was still alive and that he had won the California lottery. He gave me all the money I needed to finish my undergraduate and medical education. In the dream, I saw my father sitting in the audience watching me receive my diploma, and afterwards he kissed me and said how proud he was. Then he leaned over to whisper in my ear, "Your dream is never beyond your reach."

I spent many hours after school working various jobs. I worked as a medical and research assistant, library clerk, and chemistry tutor. Though life was challenging during my years as a college student, I never considered giving up. I was determined to stay up all night if necessary to learn the material so I could pass my classes with As and Bs in order to get accepted to medical school.

I attended study groups almost every night while I was at CSUF, and I contributed to my groups as much as possible so that we would all ace the tests together. It was fun learning science with my fellow classmates, as we were all committed to help each other do our best. There was never a moment for me in America where a teacher was bribed to move me to a higher class or where an out of control uncle screamed and yelled at me because I could not do my math homework correctly. My peers and qualified professors were my teachers and encouragers. I was pleased that I had acquired enough self-discipline to excel in my studies. I had a strong impulse to succeed and achieve, and the typical college-age priorities of having fun and dating were not going to get in my way.

I soon learned that to improve my chance of being accepted into medical school, I would have to spend hours doing special research projects as well as continue to maintain high grades. I met a lady who worked in the dean's office for the College of Natural Sciences and Mathematics. When I went to her office one day, she told me there was a great scholarship I could apply for: the McNair/Star Program, named after the first African American astronaut, who died in the space shuttle *Challenger* accident in 1986. I was eager to apply, since the deadline was just a few days away. I called people all over CSUF to get letters of recommendation and other required documents. I stayed up all night preparing my paperwork and writing my personal statement. I submitted my application close to the deadline, and within one week I heard the good news that I had been accepted as a McNair/Star Scholar at University of California, Irvine (UCI), where I would receive a stipend to do environmental research for one year. I commuted from CSUF to UCI research labs during the week and on weekends. The coordinator of the program was very friendly and helpful. She told me about other research programs I could apply for in the summer.

One of these programs was at the Berkeley Undergraduate Research Institute, and it was there that my life began to significantly change for the better.

10

Gaining Experience, Giving Care

I applied to Berkeley Undergraduate Research Institute and was accepted after a grueling application process. But all the time and effort I put into getting accepted was worth it. I met students from all over the world who showed me how to maintain my motivation as a premed student and how to study smarter not harder. Such a student who changed my life for the better was a seventeen-year-old undergraduate straight-A student from Cornell University. Right away, I noticed her energy and determination, and we soon became friends as she taught me how to study for the Medical College Admission Test (MCAT). I often visited her room late at night and found her vigorously chewing on crackers to keep herself awake as she prepared for the MCAT. After being around her, I became determined to work as hard as she did, and soon I wanted to push myself even more: Within one year, I took my MCAT.

During that life-changing summer at the Berkeley Undergraduate Research Institute, I lived in the International House (known as I-House) and ate all my meals in the cafeteria. For the first time, I ate only American food. I quickly learned how fattening pancakes, bacon, hamburgers, and pizza can be. Without my former diet of vegetables and lean protein, I started adding unwanted pounds. Despite the extra weight, I had a very productive summer working in the labs, attending lectures, making new connections, and traveling. For the first time, I traveled to Yosemite National Park with a group of French, German, and other European students from the I-House. The spectacular beauty of Yosemite took my breath away with its cascading waterfalls and brilliant rainbows of red, blue, yellow, and green. Seeing the bright colors reminded me of when I sat on my father's lap watching the Chinese New Year dragon parade right before the war erupted in my country. I felt my father's love for me all over again and thought about his profound statement that my dream would never be beyond my reach. I was living my dream—and his—working hard to become a medical doctor.

After the Berkeley summer program was over, I headed back to Southern California to finish my undergraduate work at CSUF. I started working again in the labs and doing research for my biochemistry professor on the effectiveness of the drug tamoxifen for the prevention and treatment of breast cancer.

Shortly after returning to CSUF, I was elected as a minority chairperson in the Student Health Profession

Association, a professional organization that helped undergraduate students prepare for careers in health care. This position was very rewarding because it allowed me to guide students who wanted information about what professional schools were available to them. Being a chairperson also allowed me to develop important professional connections. I was able to invite noted scholars, professors, dentists, pharmacists, optometrists, and medical doctors to speak at the student center on Fridays after school. Many of the speakers I invited were wonderful role models. One of the speakers was a kind surgeon who welcomed us to volunteer at Martin Luther King Jr.-Harbor Hospital (MLK) in Los Angeles, then known as Martin Luther King Jr./Drew Medical Center.

One night, I went to the trauma center at MLK hospital to observe him. I was very impressed with his ability to work well under severe pressure. Several patients had come into the trauma center that night with gunshot wounds. He was in charge of the surgeries, and many surgeons worked under his direction. I was impressed with their tireless efforts to save lives, but some of the wounded did not survive, as their injuries were too severe. As I witnessed these dying patients on the surgery tables, I felt paralyzed and sad. I could do nothing except helplessly watch them die. After being in the emergency trauma room for more than twelve hours, I came home the next morning at 7:00 AM and fell into bed exhausted. For a moment, I wondered if being a doctor really was my dream. I questioned whether I should be doing this work, as seeing so many patients die was very difficult and depressing. For over a week, I could not sleep or eat. I kept seeing the vivid images of those patients lying on the operating table as blood from the gushing wounds of the dying dripped on the floor. I needed answers, direction, and consolation. I sought out my friend who had been a surgeon in Cambodia, and he advised me to not listen to my fears and depression. Instead he told me, "Concentrate on learning as much as you can and focus on the positive side of medicine—saving lives." He further encouraged me by saying, "Always try your best, and never give up!"

What I saw that night at MLK hospital truly motivated me to do everything I could when given the opportunity to educate gang members about drug and alcohol prevention and treatment. It made me want to encourage troubled students to work hard to stay in school, learn all they can, graduate, and do well in their lives, because a life of drugs and violence is a dead end.

Less than a week later, I went back to MLK hospital, and the same surgeon I had observed before told me about an organization called Liga International–The Flying Doctors of Mercy, which arranges trips to provide free health care to people in need in the state of Sinaloa, Mexico. I told him that I wanted to be involved in the project and he could count on me going there soon. In a few days, I was boarding a four-passenger plane to Mexico. I was in one of thirteen planes filled with medical volunteers from Southern California.

We arrived in the town of San Blas in Sinaloa, Mexico, around 5:00 PM on a hot, dry Friday afternoon in mid June.

I was surprised to see the planes land in the middle of dry brush and a few scattered leafless trees. We immediately boarded a taxi and traveled for a few hours over rough, unpaved, dusty roads until we arrived in a desert town with run-down houses widely spaced from each other. I learned that American volunteers had built a tiny medical clinic out of one of the houses, and they had constructed a makeshift operating room. I had not seen this kind of crude setup since leaving Cambodia and the refugee camps. Hundreds of sick patients were waiting to receive our services and care, as our visit had already been promoted several months before on the radio and by word of mouth. Some of the patients had been waiting for hours before we showed up.

I remembered my experience having a tonsillectomy as a refugee in Thailand, and I felt great compassion and understanding for these people. I saw children and adults whose eyes became fresh with hope for healing when they caught a glimpse of our supplies and medicine. Some patients sat on the steps of the stairs; others relaxed on chairs or tables. Without delay, we opened our clinic and continued seeing patients until the late hours of the evening. We finally went to bed much later than our bodies preferred. We slept on the rooftop of the clinic where it was cooler, as there was neither air conditioning nor a fan available. Out on the rooftop, mosquitoes were not our friends as they circled our heads and tried to nourish themselves on our delicious blood. I was taken back to the nights I battled mosquitoes during my time under the Khmer Rouge regime and in the refugee camps. I tried to fall asleep on the rooftop with the mosquitoes buzzing around my head, but I could not. I got up and searched for a safe, mosquito-free place. I found a dentist's chair in the clinic and sat in it for a while, but my fears of falling out got the better of me, so I got up and decided that the floor next to the chair would have to be my bed. And so it was.

The morning dawned earlier than I would have liked because I was awakened by a loud vibrating sound. I thought it was an earthquake, like those I had experienced in the refugee camp in the Philippines and in California. I bolted from my sleep, fearing the worst. I peeked through the front window of the clinic, and to my amazement, I saw that the loud vibrating sound was coming from hundreds of patients hurriedly lining up outside the doorway, waiting to be seen. Soon, the rest of the volunteers came down and joined me, and we all began our work again—without attending to our personal needs, except for the necessary bathroom breaks. We did not even eat breakfast; our only meal came at dinnertime, after a very long day of work.

More surgeries were done Saturday than the previous day. By the close of the day, we still had many patients waiting to be seen. Sadly, we had to stop our work, as our hunger and exhaustion were beginning to overtake us. We did not want to risk giving medical care when our own mental faculties were severely strained.

After we closed the clinic, we all agreed that going to another city to rent a hotel room and eat a hot meal would be necessary for our well-being. We rented another taxi, and soon we were in a small, older hotel. We were all very thankful for the air conditioning, though it was loud and sounded like it might break down at any moment. We were relieved that there were no mosquitoes buzzing around our heads and that we could sleep in clean, warm beds in the rooms we shared. After we checked in to our rooms, we went downstairs to the restaurant for a late dinner. We ate hot and spicy Mexican food that was much appreciated by our very tired and hungry bodies.

On Sunday morning, we boarded our little planes again to head back to California. Our adventure as medical missionaries in Mexico had come to an end, and I would have to begin the not so welcomed adventure of studying for the dreaded MCAT.

11
Through The MCAT Hoop

To gain admission to a medical school, I would be required to take the Medical College Admission Test (MCAT). The MCAT tests many subject areas, including Biology, Chemistry, Physics, and English. My English skills were not great, and I was concerned that I would score too low to gain the attention of a top medical school. But I was determined to do whatever was necessary to score high. I had come too far to reach my dream to let less-than-perfect English skills stop me.

I learned very quickly that studying for the MCAT was tedious, repetitive, long, depressing, and at times mind-numbingly boring. I sat for hours listening to lectures, studying out of highly condensed books that were sometimes difficult to understand, taking practice exams, and listening to monotonous tapes. There were times that my brain could not absorb any more material, but I pressed on, believing that my short-term discomfort would eventually reap a long-term reward: acceptance to medical school.

Studying for the MCAT was not only difficult but also expensive. I registered at Kaplan to take a course that would prepare me for my medical entrance exams. The course cost more than I could afford. I wrote an essay to Kaplan explaining why I wanted to take this test and what my current financial situation was. Thankfully, I was granted a partial discount. Before signing up at Kaplan, I worked several jobs doing research at Berkeley and Irvine. I also had applied for loans and scholarships to pay for my extra educational expenses. I had to sacrifice some short-term pleasures to meet my expenses. I gave up spending money on new clothes, and I stopped taking pictures. Buying film and getting the photos developed was too expensive on my next-to-nothing budget.

One of my heroes and inspirations during my MCAT preparation days is now a top anesthesiologist in Korea, Dr. Mijeung Gwak. I met Dr. Gwak at Kaplan, when she was studying for medical board exams. Every day, she brought food and graciously allowed me to eat her nutritious Korean food. She also provided me with excellent advice about the MCAT exam and shared great tips on how to ace the test. I learned from her that successful test-taking means understanding how to take a test. Knowledge of the subject matter is one part of succeeding on a test, but pacing oneself, maintaining mental stamina, and developing a strategy are also critical to success on exams.

Another role model during this time was my high school friend, Dareen, who now practices medicine and specializes in treating allergies. She provided wonderful moral support, good advice on what medical schools to apply to, and insights on how to prepare for the medical school interview.

I visited the clinic of Southern California dermatologist Dr. Marina Ball when I was bored with studying for the MCAT. Her depth of knowledge about dermatology, gentle guidance, and direction helped me tremendously. She showed me the roots and causes of dermatological disease before I entered medical school. She was always eager to help me in whatever way she could, as she was excited that I had chosen a career in medicine.

The day finally came for my MCAT exam. I couldn't sleep the night before. I felt anxious and scared, as this one exam would decide my future direction: Would I be headed for medical school or would I have to try again?

The exam lasted for almost ten hours. I remember how relieved I felt after I finished. I walked out of the room feeling like a limp, over-boiled spaghetti noodle. I would have to wait two to three months to receive my MCAT results. During the waiting period, I volunteered at various medical clinics, continued doing research, and concentrated on finishing my last semester at CSUF. I went to my mailbox one day and found that my MCAT scores had arrived. With shaky hands, I ripped open the envelope. I was surprised how well I did in physics and chemistry. I was pleased with my test results, and I could now apply with confidence to various medical schools and start writing personal statements, collecting recommendation letters, and obtaining the other documents required for acceptance to medical school. I worked many hours preparing the medical school application packages. Sometimes I was up very late at night writing my personal statement essays. I often fell asleep, but I soon woke up and continued working, believing that I was just a step or two away from my dream.

I sent out many medical school applications, as I knew how competitive the admissions standards are for getting into a top-rated medical school. By Thanksgiving that year, I had been accepted to go to medical school in California. I was very excited, and I had the best Thanksgiving the year that I got accepted. I also enjoyed Christmas even more, knowing that I was definitely going to medical school. But I was still waiting to hear from Howard University. One day, shortly after returning from an all-expense-paid trip to France—a very special congratulatory gift from Uncle Heng's daughter for getting into medical school—I received a letter from Howard University, College of Medicine. I eagerly ripped it open and read: "It is with great honor that the Family of Howard University, College of Medicine, has accepted you as an entering freshman." I jumped up and down and screamed. I was the happiest person on earth. I was on the top of the world. My dream of becoming a medical doctor was about to come true, and my father was right. But my struggles to succeed were not yet over.

12

The Doctor To Be

I moved to Washington DC seventeen days before starting medical school. I could hardly wait. The thought of starting school was very exciting, and I was looking forward to finding out who my roommate would be. My roommate turned out to be a hard working girl from Nigeria who was also one of my classmates.

When I arrived at Baltimore/Washington International Airport, I soon discovered at the baggage claim that all my suitcases had been lost, including my bedding and clothing. Everything was gone except an anatomy atlas that I had carried with me on the plane. I was disappointed over my loss, but I was focused on what was really important to me—succeeding in medical school. If I had to wait for my bedding and clothes, I could do that. I ended up sleeping on the carpet in my room for some time without bed sheets or coverings of any kind.

The airline that lost my luggage did not deliver it until several days later. I had no money to buy extra clothes, so I wore the same outfit day and night. I didn't go out except to buy food, and for hours every day I studied anatomy. I never allowed myself to be ungrateful because I knew that I was rich compared to the extreme poverty I lived in as a child growing up under the Khmer Rouge regime and in refugee camps. I was happy to be alive, to have food to eat, and to sleep in a nice, clean apartment with running water and toilets that flushed. I must say that I was extremely happy to live without the annoying buzz of mosquitoes around so I could finish my anatomy atlas before medical school started. I knew that learning anatomy would be a challenge, and I didn't want to waste a moment of time.

The first few weeks before school started were difficult. I was lonely. I was in a strange new city, and I missed my family, friends, and especially a young man I met in chemistry class who had moved to Hawaii. Whenever I felt lonely, I went to the computer lab and emailed my friends. Once school began, my loneliness faded as upperclassmen welcomed me right away and invited me to be part of their study groups.

At Howard University's College of Medicine, the freshman class consisted of students from wonderful diverse cultures from all over the world. I was pleased that all students eagerly accepted one another. Many students became instant friends, and there was no racial tension. We were too busy studying to think about discrimination and too busy helping each other succeed. We had the common goal of graduating with our medical degrees and pursuing successful careers as doctors. Together, we would be strong. As the Chinese proverb says, "A cord of many strands is not easily broken."

On the first day of class, I met a woman who would become one of my best friends, Aretha Makia. We met in the restroom and immediately struck up a conversation. From then on, we were inseparable like twin sisters. I saw that she was expecting, and I admired her courage to enter medical school pregnant. Her husband was already practicing medicine in a private setting. Aretha told me that she was a former Miss Cameroon. I was not surprised, as she was very pretty, tall, and thin. I saw many gorgeous photos of her in beauty pageants and on modeling photo shoots. I respected her decision to become a doctor and give up a life of beauty and ease to help others—and for her it meant commuting two to three hours a day in heavy traffic, which resulted in loss of sleep.

The first exams in medical school were mentally taxing. I did not know how I would make it, yet I persevered with encouragement from other classmates. We all agreed that the classes were too long, and there was just more information presented than we could put into our brains in one sitting. Night and day, we went to class together, ate together, studied together, and repeated this routine for many weeks. Taking a day off was almost never done, unless a student was very sick. In my first year, I had a severe case of bronchitis and still crawled to class, knowing that if I missed one day I would have to work really hard to catch up with the very high volume of materials we were plowing through: biochemistry, anatomy lecture and lab, pharmacology, microbiology, physiology, and more.

Receiving the results of our first anatomy exams was upsetting. We all wanted to pass, but not everyone did well. Those who did not pass the exam cried, but we hung together and encouraged each other to stay the course because we all knew that we were only truly successful if we were all successful. Dr. Wilson and Dr. Aziz were two of our anatomy professors; they gave up their weekends, holidays, and nights to help us learn the volumes of materials we had to know to pass the exams. Dr. Wilson told our class, "Girls, please do not date now. It would be too distracting. Your boy toy will come after you graduate." It was wise advice.

During that first year, I went out with my single classmates as a group after every major exam to relax my brain. Money was tight, but that did not stop any of us from having fun. We would laugh and enjoy being together as we happily carpooled in a big car on outings to Dupont Circle, Adams Morgan, or the Waterfront along the Potomac in Washington DC.

Before long, I experienced my first East Coast winter. I noticed the changing seasons as I was looking out my apartment window every morning. I could not escape the startling beauty of the leaves turning bright yellow and orange. I asked my roommate when all the leaves would be off the trees, and she said, "Soon."

It was as if the changing of seasons paralleled our lives as new medical students. We started out fresh and green, and then slowly began turning to deeper colors as we matured in our understanding of the materials and in our relationships with one another.

In the early months of our first year, we learned anatomy and we were up late, sometimes until 2:00 AM, dissecting cadavers and learning the many muscles, arteries, veins, nerves, and bones of the human body in a very short time. Sometimes our brains would get numb trying to absorb all the new information. One early morning, while I was driving back home from another late night of studying, I noticed how cold and dark it had become and how slippery the roads were. I did not have chains on my tires, and I was afraid I might not make it home safely. I was not used to driving on icy roads. My car swayed very rapidly for many minutes. It was so hard to control the wheel in such icy conditions. I wondered if my life might end in an accident or if I would hit the big trees near the road. I began to pray, knowing that the ice on the road was not the only danger—there was also the possibility of meeting a drunk driver head on as many left bars late at night intoxicated. I prayed very hard to not run into problems that would destroy my life and my dream. My prayer was answered, as I arrived safely at my apartment and started to study again. I finally went to bed and only slept one hour before my alarm went off loudly. Startled, I shook with fear when the alarm sounded. I managed to wake up, and I was back in anatomy class at 8:00 AM.

Often I did not eat breakfast or lunch because I stayed in class extra hours to learn the materials. I addressed my hunger by packing snacks to eat in class and also eating during small breaks. Our brains needed glucose to be able to absorb all the new information we were learning. So we munched during class, and thankfully, the professors did not mind.

I drew on God's strength constantly during my medical school journey. I prayed my way through exams and long study sessions. I prayed for my fellow classmates, too, because I was committed to their success as much as my own.

Sometimes loneliness would take its toll on me, since I did not have anyone I was close to like a boyfriend. But we were all in the same boat, struggling to make it through and achieve our dreams. On occasion, I tried reaching out to my relatives in California. Instead of receiving encouragement when I called them, however, I was blamed for going to medical school three thousand miles away and told that I shouldn't complain because my loneliness was my fault. They blamed me for not going to one of the medical schools in California that accepted me. This was not what I needed to hear. The berating continued, and one relative went so far as to say that I "might not pass medical school." I hung up the phone feeling worse. Now I was lonely as well as depressed. But I chose not to listen to the words of discouragement. Instead, I focused even more on getting through medical school.

The next day, I went to the library very early, determined that I would study harder—not only to prove my relative wrong but also to get a lot of my studying over with so I could relax and go back to California during Christmas break, which was only two months away.

Besides the extra stress I was putting myself through by studying longer hours, money worries were a constant. There were times when I worried whether I would have enough money for tuition. I had a fear of being kicked out of school for nonpayment of fees, and it was this concern that sometimes kept me awake at night. I comforted myself by remembering that I had survived far worse circumstances and made it through. I thought: *Why should these difficult financial times at medical school be any different?* I put my worries aside and pressed forward, still believing very strongly that my dream was closer to being within my reach than it had ever been before.

Days and weeks went by very quickly, and I noticed that by the end of November, almost all the yellow and orange leaves that decorated the trees were on the road, leaving the trees bare and bony like the skeletons I looked at for hours in the anatomy labs. Looking at the bare trees amplified my loneliness, and California felt even farther away. I missed the warm, sunny days; the sandy beaches; and my friends and family. I began to literally count the days—forty-five, forty-four—all the way down to the very day I took my final exam. As soon as the last exam was over, I was on a plane heading west, homeward bound.

My sunny California beach days only lasted two weeks. I did not take much of a break from studying. Soon after I arrived, my books were open again because I knew it was only a matter of weeks before the second semester would start, and I had to take a Neuroscience course that would require me to understand large amounts of complex information. Advanced preparation would be important for me to pass the class.

When my second semester began, I discovered that it was significantly more difficult than my first semester. My class load nearly doubled. I spent hours in one of my classes looking at histology slides. In the beginning, I was bored listening to long lectures that almost put me to sleep. As my knowledge of the subject area grew, I discovered that the boring lectures were quite interesting. I knew I had to pass my classes so I could leave feeling confident that spending my summer working in California would be a good use of my time. I did not want to have to stay in DC retaking the course if I failed to pass.

The week before finals, I slept for only two hours each night. My classmates and I were in the library until 3:00 AM hunching over our books, memorizing our lecture notes, and cramming our brains with detailed information that covered almost everything we learned in our first year of medical school. One night after a library study session, I got into my car to return home, and my old car engine died. Nothing I tried could get it started again. I was not in a safe area; many thugs lived where I found myself stranded. Thoughts of Cambodia and the bandits that tried to prevent us from leaving entered my mind. I was afraid. I used my cell phone to call my friend Aretha. She quickly answered her cell phone and soon came to my rescue. She took me to my apartment, and then just a couple hours later, she came back and picked me up. Once again, we were off on our study adventures, determined to pass our first year of medical school.

The following fall, after a paid summer fellowship at Childrens Hospital Los Angeles, I would begin my second year. I appreciated more than ever the value of having classmates like Aretha and others who provided encouragement and inspiration as the course work became increasingly difficult.

13
Rotating Through the Grind

My second year of medical school began with a new concern. I learned that I was required to take a very difficult exam, the USMLE 1, one of the first three board exams medical students must pass to be able to practice medicine in the United States. The pressure to pass was extremely high, as every student wanted to move on to the third year and be one step closer to achieving their goal of successfully completing medical school. I again focused harder on my studies, trying not to allow any unnecessary distractions. The class load increased again, and the amount of time spent in each class was longer. I had a block of four hours of pathology lectures two to three days a week, and then it was off to the labs for additional study and research.

I often wondered how I was able to remember the volumes of information I needed to know to pass pathology as well as my other classes when I was getting so little sleep. I did not really know how I absorbed the dozens of books and lecture notes to successfully pass the exams. It was as much of a miracle to me as leaving Cambodia without being shot and killed.

Despite my tired body, I enjoyed reading my pathology books. I particularly loved pathology slides, and I spent many hours a day looking at them, sometimes missing lunch and dinner. I was fascinated by the beautiful colors and intricate patterns of each slide. I was in another world, and a small microscope was my window.

With so much study and processing of very dense information, there were times when I felt my brain shut down; like an over-soaked sponge, I could not hold any more. During Christmas break in my second year, I went to visit relatives in St. Thomas in Ontario, Canada. I relaxed for a time, but soon began reading pathology textbooks again during my break. My next challenge would be the first of three medical boards that I had to pass; and studying was what I needed to do to succeed. A vacation was not going to stop me from putting my mind on my board material. The board was far too important.

After I took the first board in July, I flew to California to work on my fellowship with the American Academy of Dermatology. I was not worried about my board results because I had studied well and was confident that I knew the information I was tested on. During my time working in dermatology, I received a stipend as I worked in research and saw patients with various skin diseases. I learned how to quickly diagnose common skin diseases as well as rare ones with the guidance of excellent attending professors. I worked at several hospitals, including UC Irvine Medical Center, VA Long Beach Medical Center, and the Beckman Laser Institute in Irvine.

Shortly before beginning my third year, I learned that I had passed my first board. What awaited me next were clinical rotations that my friends warned me would be very difficult to pass. My first rotation was obstetrics/gynecology (OB/GYN). The first day of the rotation, I was to be on call all night with other students. We were not told the schedule ahead of time, so when we got the call that we were expected to be at the labor and delivery ward right away, we were not allowed to go home and retrieve our personal supplies. Any student who tried to do so would have their grade negatively affected.

It was 10:00 PM when we arrived at labor and delivery. It was not busy, so we were told to go and work in the operating room or get some rest. We chose to rest, as our bodies were exhausted. But as soon as I put my head on the pillow, I heard the sound of wheels rolling across the floor. A stretcher was on its way, and a pregnant woman needed an emergency caesarian section. I had to get up and report to the operating room (OR) to be in attendance at the emergency surgery. It was there that I saw a baby delivered via c-section for the first time. Just a few seconds later, a distraught pregnant woman in her third trimester with painless vaginal bleeding and a placenta previa diagnosis was rushed into another emergency room, and we had to scrub our hands, arms, and fingernails and put on fresh gowns to help with this new emergency. After the two emergencies in OB/GYN, I made it back to bed at 5:00 AM and could only sleep for an hour as I had to be up again at 6:00 AM to prepare for morning rounds to help decide which therapeutic regime would be best for each individual patient.

Once the morning rounds were completed, my work was still not over. I worked with the residents, furthering patient relations, checking charts, providing reports, documenting cases, and doing more patient care. At times I felt like a zombie, but I forced myself to stay alert. I knew the important position I was in, as I was offering help when lives were in the balance. After several weeks of a very rigorous routine and schedule, my OB/GYN rotation was finally over. I still had to take a national pre-board exam, which I successfully passed. It was on to the next rotation.

My next rotation was general surgery. It was surprisingly more painful than OB/GYN. General surgery rotation at Howard was an unforgettable experience. Gunshot and car accident victims frequented the emergency room. Blood poured from victims' wounds, and sadly, though everything possible was done to try and save them, many of them died. This was hard on me, but not as difficult as my first experience with dying patients while volunteering at MLK hospital in Los Angeles. I understood how my volunteer experience had prepared me for this general surgery rotation.

For my general surgery rotation, I had to report for work at 4:00 AM. I did not care about putting on makeup or nice clothes. I knew that I would be in the OR for several emergency surgeries, and I would not come home until after 10:00 PM. Once I arrived at the hospital, I made my patient rounds, checking vitals and doing whatever was necessary to ensure the well-being of the patients I was responsible for.

Even though I had to keep long hours during my various rotations and suffered from sleep deprivation, I still had to study for my next board, USMLE II, and I spent hours in the library doing research on my patients' specific illnesses as well as preparing a personal statement and requesting letters of recommendation. I had almost more than I could handle during my third year, and then I had to take my pre-board exams in OB/GYN, surgery, neurology, medicine, psychiatry, family medicine, and pediatrics—and each one was a two-hour exam. The pre-board exams were designed to prepare us for USLME II, but I did not think I (or any other medical student) needed an exam to prepare for an even bigger exam. I was so very tired of taking exams, especially when my body constantly ached for sleep.

One day when I was especially tired and working with a young, healthy, married, male patient, I accidently stuck my finger with a needle that I had just used to draw blood from him. He had come into the clinic as a stable asthma patient but needed complete physical exams. Thus, I drew his blood. After I pricked myself, I worried that I might contract a blood-borne disease, even though the likelihood of that was slim if I went through the necessary procedure of taking prophylactic medications. I knew exactly what I needed to do to avoid getting any blood-borne disease the patient might have had.

From the moment I pricked myself, my instincts and knowledge about what to do kicked in. I immediately ran over to the ER at Howard University Hospital, which was only about one hundred yards from the clinic where I had been working. Several health-care staff members drew my blood, and I was given a combination of prophylactic medications in less than a four-hour period from the time I was stuck with the needle. I had to take these prophylactic medications for one month. During this one-month period, I feared that I might have caught some kinds of disease; but I prayed every night that my blood would be clean. I was retested after the month passed, and much to my relief, the tests came out negative.

14

Ready to Practice

I passed my third year and moved on to my fourth year where many more rotations were ahead for me. One of my favorite rotations was general surgery during my senior year. I worked with a surgery resident, Dr. Fripp, who is now a top plastic surgeon in Georgia. She encouraged and inspired me many times to reach my goals. Shortly after I passed my USMLE II board exam, she proudly congratulated me, saying, "Sophie, you are now one of us."

In the beginning of my fourth year, I worked on the poster presentation for the American Academy of Dermatology (AAD) at its National Annual Meeting and Conference in Washington DC in early March on a drug used in acne treatment. I flew to DC to present my poster and attended many other symposia and lectures on skin health for one week. There were so many outstanding posters and lectures from all over the world.

After the poster presentation, I flew back to California. I was so excited to be there for four months. I spent a lot of time interviewing for residencies, and I worried that I might not find a residency program in Southern California. But my concern was ill founded, as I matched at one of the places I preferred: LAC-USC medical center.

After senior general surgery, I went to the internal medicine rotation, where I saw very sick patients with incurable conditions, including AIDS, end-stage liver disease, end-stage renal disease, and other terminal cases of chronic disease. It was beyond our ability to help these patients because they were too sick by the time they came to the hospital. I remember running around the hospital with my residents to help our dying patients. Due to the severity of the diseases and the high number of patients I saw during this period, the good that came from this time was that I developed a strong foundation in being able to successfully diagnose and treat patients with acute and difficult pathologies.

Senior medicine was only four weeks long, and it would be over very quickly. When it was over, I flew back to California to start a dermatology elective at MLK hospital in Los Angeles. It was my first dermatology rotation that I assisted in dermatological-surgical procedures. In the middle of the rotation in mid December, I was invited by the AAD to go to Capitol Hill for the congressional hearing of what was then the most potent and controversial medication used in the treatment of severe and recalcitrant cystic acnes. I sat in the middle row with other professionals for many hours, praying that I would not be called to the front to testify about the quality and side effects of the medication. I was only there as a favor to my friend who worked for AAD. I was happy my prayer was answered and I was never called on. After sitting in the room hearing testimony for hours, I was very relieved to get on an airplane from DC back to California to finish my dermatology rotation at MLK hospital.

At MLK hospital, my chief attending professor was A. Paul Kelly, MD, who taught me a lot of dermatological procedures by having me work closely with him in the clinic as he treated patients with rare and unusual conditions. Usually, these cases are only seen in medical textbooks. I learned so much from this excellent doctor, though I only worked with him for four weeks. He encouraged me, as well as other students and residents, to go to Metro Dermatology Conferences that were held to discuss difficult skin cases that a solo doctor cannot treat. In addition to working with Dr. Kelly and attending various conferences, one of the dermatology residents gave me a book called *Principles of Dermatology*. I stayed up late reading the book every night. When I finished the book, my dermatology rotation had ended but my love of learning in the field of dermatology continued. I took one more elective course in dermatology at UCLA. It was there that I met another great mentor and professor, Gary Lask, MD, who inspired me with his knowledge and passion for dermatology.

After completing my dermatology rotation, I returned to Washington DC to find out where my residency was going to be; I had been hoping that my match would be in California. I got to school late on my first day back, due to heavy traffic in DC that morning. I rode with Aretha and her two toddler sons. They were as excited as we were to celebrate our victory, as they had been born during the years we were in medical school. We both were delighted to receive our letter from the dean that gave us each what we desired. I was going to be a resident at Los Angeles County-University of Southern California (LAC-USC) Medical Center, and Aretha was going to be at a hospital near where she lived in the Baltimore area.

Later that afternoon, we attended a wonderful reception given to us by the medical school. The reception was in the same room where we had taken almost all of our grueling exams. The place of struggle became a place of joy and sweet victory. My dream was no longer in doubt—it was a reality that I was sharing with other medical students that jubilant afternoon. We further celebrated our victory with dinner out at Union Station. Not one of us could believe that our hard work and perseverance was now paying off. We were living our dream. That evening after dinner, we had a candle light session on stage at our school. Everyone held a lit candle while holding hands and singing songs. At that moment, I reflected back on my father's dream and remembered once again sitting on his lap as a young five year old watching the Chinese New Year dragon parade with all the flashing lights full of bright hope and promise. My strong hope had been fulfilled, and my sometimes bittersweet dream had finally arrived. As I held the burning candle above my eyes, I had a feeling that my father and Uncle Heng were watching from above and were very proud of me. I believed that my father knew the seeds he planted in my heart at such a tender age had sprouted, become full blooms, and would one day bear much fruit. Tears rolled off my cheeks as I stood with my fellow classmates, feeling extremely grateful that we had achieved our dreams. Indeed, together we were better.

In the next few days, I would go to California to finish another dermatology rotation and then return to DC for one more month to start my senior family practice rotation, which would be my very last rotation as a medical student. I went back to DC and stayed with Aretha and her family again. The morning before I left, I made breakfast for her two sons. They knew that their mom was going to be leaving soon, and they started to cry. I almost cried too, as it was very hard to leave the boys. I had become very much attached to them, having bonded with them before they were born. I knew them while they were still babies in their mother's womb and had watched them grow up into healthy and happy toddlers. Both Aretha and I left that morning and returned later that evening. I'll never forget how the boys peeped through the upstairs windows and saw us pull up to the house; they ran down the stairs, flung open the door, and flooded us with hugs and kisses. Moments like that could not get any better.

Having received the news about my residency, for the first time in a very long while, I could read books that were not medical textbooks. I started reading about personal relationships and finished the books I had in only two days.

I wanted to be married, now that graduation from medical school was just around the corner—three weeks, four days, two hours. I could not wait.

I flew back to California thinking that the time might pass more quickly. I also had to do more rotations to meet the last requirement before I graduated. But time did not pass as quickly as I'd hoped because I was lonely. I did not have the man I loved. We had kept in correspondence via emails and very frequent phone calls during medical school. I liked him because, like my father and Uncle Heng, he had a good and kind heart. We shared our lives as close friends; I longed for more. But our relationship turning into anything deeper seemed like a distant possibility, since he was living in Honolulu, Hawaii.

15

Sopheap Ly, MD

I woke up the morning of my graduation at Aretha's home, feeling tingles of excitement. My dream was coming true. I hurriedly dressed, and soon Aretha joined me. We hopped into her car to go to the auditorium at Howard University's College of Medicine, which was about an hour from her house. The Honor and Oath ceremony started at 10:00 AM, and we were stuck in rush hour traffic. We worried that we would be late and that we might miss graduation. But our fears were not founded, and we made it to the auditorium in time to quickly put on our caps and gowns. We joined our other classmates, who shouted, "Sophie, Sophie, Sophie!" Like me, they wondered if I was going to arrive on time and make the lineup. Thankfully I had, so both Aretha and I quickly got in line, and instantly we could not stop smiling. How could we help it—all of us were living our dreams! Together, we had struggled to stay awake during marathon study sessions and also endured countless sleepless nights, mind-cracking exams, and depressing rotations where death sometimes reigned. Now our moment of victory had come. How proud our professors and deans were of us, and it showed as they brightly beamed from the podium.

As we marched to the main auditorium, I could hear the crowd cheering, and I knew that my cousin from France, Uncle Heng's daughter, and other relatives from California had come to support me on my dream day. I knew that my father and Uncle Heng were there cheering, too, from their seats high above.

Before I sat down, I saw something that made adrenaline surge through my veins. There on my seat was a large black leather bag with gleaming letters: "National Medical Association." In a matter of minutes, I would be granted a medical degree, and I would soon have after my name, Sopheap Ly, *MD*. My imagination soared, and thoughts raced through my mind: *I'm going to be a doctor*. I flashed on a mental image of a stethoscope hanging outside of a barbed wire fence. I was finally out of my refugee camp.

I heard my name called, "Sopheap Ly," but it was the word *doctor* before it that thrilled my heart. I received my "diploma" on stage, and shook the dean's hand just like I dreamed I would years before. I took my seat again and watched the crowd of cheering faces, and before long, all my supporters were hugging and congratulating me.

My graduation day was soon over, but there was one unanswered question that lingered in my mind: Was my blood still clean after I had pricked myself accidently over a year ago? If so, I could move on with confidence that I would be practicing medicine, but if not, then my years of study and sacrifice might be put on hold until I was treated for a disease, should one show up in my blood work. I had to know the truth. Leaving the crowd and happy smiles behind, I removed my cap and gown and raced to the employee health center to get my blood drawn and retested. As the nurse pricked me, I prayed that my test would come out negative. I had to wait a few days before I found out the results. But I did not wait for a letter to arrive in the mail. I called the lab many times and did not stop calling until I heard the news: "Sophie, your test results are negative." To say I was glad would be an understatement—I was ecstatic.

With my blood clear and my mind put to rest, I was ready to receive my real diploma. What was handed to us on stage during the graduation ceremony was only a mock diploma. As I stood in line waiting to receive the real thing, I could not help thinking how blessed I was to be a part of such an awesome class of medical graduates and to have been taught by some of the finest professors in the country. My heart was filled with gratitude, and I gave a big thank you to God. My dream had proved to be within my reach.

16
Pearls of Wisdom

Gratitude. Wherever you find yourself at a given moment—be it on stage graduating from medical school or in a bamboo leaf hut in a refugee camp—there is always something to be thankful for, even if it is just that you're alive and breathing.

America. America is a wonderful place. Never take it for granted. In the United States, anyone can choose to work hard and make use of the opportunities that are available. I did, and I was only a poor refugee from a war-torn country who had nothing but the clothes on my back when I first stepped foot on American soil.

Courage. Only the bold and courageous dream. Everyone can dream. Courage is there for the taking if you choose to press on despite the temptation to quit.

Achievement. It is not how you start out in life, but how you finish. To finish well is to dream big. Dream to make a difference. Never give up, for in not giving up you will find courage and strength to pursue your dreams.

Forgiveness. Forgiveness sets you free. Forgiveness does not mean that you forget. It means that you let go of anger toward whoever has hurt you. Holding a grudge is like tying yourself up with a big rope—you will miss out on positive opportunities in the future. Had I held on to my anger against the Khmer Rouge for killing my father and Uncle Heng and causing my family and me so much pain, I would have not been able to achieve my dream of becoming a medical doctor. I would have been too busy projecting my pain and anger onto those around me. Instead, I learned to give back and do well by becoming a part of the healing process.

Hardships. Embrace and accept hardships as your life's lessons. My suffering as a child taught me the value of courage and perseverance, and it gave me the gift of being thankful for everything. It also taught me to be kind, patient, and generous to those in need.

Opportunity. When opportunity comes, accept it, chase it, and treasure it. Do not wait until tomorrow. Take the steps to live your dream today, and run the race to win. Know that no matter how dark your circumstance is, truth and hope are always brighter. Never give up, because your day will dawn. Your dream is never beyond your reach—but you do have to reach.

Money. It's true that money and power can't buy love. Having a dozen diplomas hanging on your walls can't buy love, either. Money also can't buy good character. Good character and integrity are more valuable than all the money in the world.

Medicine. There is great value in enjoying life every day and laughing with those you love. It is the best medicine.

Health. To be healthy is to look on the bright side of life. The cup is half full, not half empty. To think happy thoughts is to be very wealthy.

Simplicity. The best life is a simple one, uncomplicated with cares and filled with childlike trust and the belief that things will work out for good if you search for truth and understanding.

Love. Receiving love from family and good friends is more important than making lots of money in your career. A life without love is no life at all. Good relationships take work, and difficult ones take even more. To love someone is to forgive.

Romance. It turns out, nothing is sexier than a man with a sincere heart. I married my close friend, the one with "the good and kind heart." The one I thought seemed like a distant possibility. We are happily raising beautiful twin daughters together.

Biographical Statement

Dr. Sopheap Ly completed her residency in internal medicine at Los Angeles County-University of Southern California Medical Center and is a certified Diplomate of the American Board of Internal Medicine. She is happily married to her best friend, Kaustubh K. Marathe, DDS, and they are the proud parents of twin girls born in 2008. Dr. Ly lives and practices internal medicine in Southern California, where she focuses on helping veterans live healthy lives.